SCIENCE AND TECHNOLOGY IN CHINESE CIVILIZATION

中華科技史文集

程[illegible]題

SCIENCE AND TECHNOLOGY

EDITED BY **CHENG-YIH CHEN** 程貞一

with Assistant Editors

Roger Cliff

Kuei-Mei Chen 程桂梅

IN CHINESE CIVILIZATION

With contributions by
Cheng-Yih Chen 程貞一
Tsun Ko 柯俊
Di Li 李迪
Zhao-Hua Li 李兆華
Ko Wei Lih 李國偉
Jing-Yan Lu 陸敬嚴
Rong-Zhao Mei 梅榮照
Shu-Qin Qi 齊書勤
Fang-Toh Sun 孫方鐸
Jin-Guang Wang 王錦光
Ze-Zong Xi 席澤宗
You-Liang Zha 查有樑

World Scientific

Published by

World Scientific Publishing Co Pte Ltd
P. O. Box 128, Farrer Road, Singapore 9128.

Library of Congress Cataloging-in-Publication data is available.

SCIENCE AND TECHNOLOGY IN CHINESE CIVILIZATION

ISBN 9971-50-192-9

Printed in Singapore by Kyodo-Shing Loong Printing Industries Pte Ltd.

CONTENTS

Science

Technology

Editor's Note

Some comment should be given with regard to the conventions used in the papers of this book. The general rule which has been followed by the presentation of Chinese technical terms, titles, and quotations is that on the first appearance of a Chinese term it is given in romanized form followed by Chinese characters. Following this is the English translation, enclosed in parentheses. Thereafter, the term will generally be represented by only the romanized form. An exception is made for long quotations for which the romanized form is omitted and only the Chinese characters and English translation are provided. The romanization system used is either Hanyu Pinyin* 漢語拼音 or Wade-Giles, depending on the choice of the author. Where a particular term is romanized under more than one system, both romanizations appear in the index and cross-reference each other.

Book titles are underlined in their romanized and translated form, and enclosed within double angle brackets 《 》 in their Chinese character form. Chapter or article titles are indicated by single quotation marks in their romanized and translated form, and enclosed within single angle brackets 〈 〉 in their Chinese character form. Chinese technical terms (or other foreign terms) as well as chapters and article titles in Western languages are indicated by single quotation marks as are their translations. Quotations if not offset by indentation are indicated by double quotation marks. The Chinese characters are, so far as possible, given in their original

*A slight variant on the Hanyu Pinyin system is used in which the associated components of a compound word are connected by a hyphen rather than concatenated into a single romanized word as is commonly done.

full form. For standardization purposes, the translations of book titles follow mainly those used by Joseph Needham in his Science and Civilisation in China, with exceptions made where there is reason to believe a more accurate or appropriate translation is possible, or where there is a specific request by the author.

Chinese personal names in the papers of this book are normally written according to the Chinese convention — family name first followed by given name(s) both in romanized form and in Chinese characters form. However, the names of the authors of these papers in their romanized form are given in the Western convention — given name(s) first followed by the family name; the Chinese characters are still given in the traditional Chinese order. Since many of the authors already have a particular romanization of their names to which they are accustomed and by which they are recognized in the scientific community, the romanization of the personal names of the authors of these papers is given according to individual preference and may or may not agree with the system used within the author's paper.

Throughout these papers a modified - and + convention for indicating dates is adopted. In this modified form the + sign is omitted except in cases where possible ambiguity could arise (e.g., during the Han 漢 dynasty). It must be remembered that astronomers, who use a similar system, insert a year 0. The minus sign dates in this book are the historical dates, not the astronomical ones.

Finally, the editors would also like to make a few remarks regarding the romanization of Chinese terms in general. As can be seen from the above description, the system used in this volume is quite cumbersome. It is the feeling of the editors that, when the capability to reproduce Chinese characters exists, the use of romanizations becomes redundant, and that in any case romanizations are of limited use. In the absence of any indication of tonal movement, romanizations are at best a misleading guide to pronunciation, especially considering their various idiosyncrasies. Although the information on pronunciation provided by Chinese characters (the majority of Chinese characters are logograms and thus contain a phonetic component as well as a signific component) is not useful for the non-Chinese reader, the Chinese characters permit direct graphic distinction of homophones. It is the inability of romanizations to distinguish homophones that gives rise to the difficulty in interpreting unfamiliar terms and titles given in romanized form. With the recent progress in computer and printing techniques, it is not unrealistic to expect that Chinese characters be reproduced within texts, thus eliminating the need for an extensive dependence on romanization systems.

CONTRIBUTORS

Cheng-Yih (Joseph) Chen 程貞一

Professor of Physics, University of California at San Diego; Director, Wei-Kung 爲公 Institute

Tsun Ko 柯俊

Professor of Metal Physics, Beijing 北京 University of Iron and Steel Technology; President, Chinese Society for History of Science and Technology

Di Li 李迪

Associate Professor and Director, Institute for the History of Science, Inner Mongolia Normal University, Huhehot 呼和浩特

Zhao-Hua Li 李兆華

Lecturer of Mathematics, Tianjin 天津 Normal University

Ko-Wei Lih 李國偉

Deputy Director, The Institute of Mathematics, Academia Sinica (Taipei 台北); Research Fellow, The Institute of Information Science, Academia Sinica (Taipei)

Jing-Yan Lu 陸敬嚴

Associate Professor of Engineering, Tongji 同濟 University, Shanghai 上海

Rong-Zhao Mei 梅榮照

Associate Professor of History of Mathematics, Institute for History of Natural Science, Academia Sinica (Beijing 北京)

Shu-Qin Qi 齊書勤

Director, Shanxi 山西 Provincial Bureau of Seismology, Taiyuan 太原

Fong-Toh Sun 孫方鐸

Professor Emeritus, National Tsing Hua 清華 University, Hsinchu 新竹

Jin-Guang Wang 王錦光

Professor of Physics and History of Science, Hangzhou 杭州 University

Ze-Zong Xi 席澤宗

Director and Professor of History of Astronomy, Institute for the History of Natural Science, Academia Sinica (Beijing)

You-Liang Zha 查有樑

Associate Director, Institute of Dialectics of Nature, Academy of Social Science of Sichuan 四川

Preface

The contribution of Chinese civilization to the development of science and technology has in recent decades begun to receive due recognition as an integral part of the world history of science and technology. This changing attitude has been brought about in part by the fact that science and technology is now widely recognized as a product of a global human civilization, and hence, syntheses based on a single lineage for the development of science and technology are logically defective and factually incorrect. Among other factors that have also contributed to this change is the abundance of Chinese records on science and technology recently made available to historians around the world through the work of a few dedicated scholars in the West and East.

With this changing attitude, the study of the Chinese history of science and technology is now entering a period of great progress as we see the near completion of the first major synthesis of the entire spectrum of science and technology in China by Needham and his collaborators and also the great accumulation of Chinese archeological discoveries in recent decades, poised for major breakthroughs in many issues confronting scholars in this field. In addition, there have been, in recent years, renewed efforts in the systematic collection and categorization of Chinese historical records on science and technology and in the organization of meetings and conferences in this field.

This book is a representative sample of current work in the history of science and technology by Chinese scholars. It is a collection of papers presented in the summer of 1985: nine papers at the San Diego workshop on the History of Science andTechnology in Chinese Civilization and five papers at the XVIIth International Congress of

History of Science in Berkeley. The value of this collection stems not only from the importance of the individual topics presented but also from the wide range of approaches, methods and viewpoints therein. There are a total of six papers on the history of mathematics, five on science and three on technology.

Instead of providing a brief commentary on each of the papers as is often done in a preface of this nature, I will make some comments on questions raised at the San Diego workshop and leave each paper to speak for itself.

A general topic discussed repeatedly among the participants and some of the speakers at the workshop can be phrased by the famous question, 'Why did modern science not first develop in China?' of which much has already been written by leading hsitorians of science such as Needham,[1] Sivin[2] and Graham.[3] My interest here is not to provide another essay on this topic but to comment on the general perceptions of this question.

In common terms, science deals with the discovery and interpretation of the truths of nature while technology deals with the invention and implementation of the machinery of man. Although the two are closely related, they are fundamentally different. In examining the question of the scientific revolution and the development of modern science, one must keep in mind the difference. Factors that are influential to the industrialization of technology or even to the technological invention process may not necessarily be influential to the process of scientific discovery. In the examination of a scientific discovery, one must realize that the discovery process cannot be precisely pre-programmed since one cannot specify what is to be discovered prior to its discovery. There is very little predictability in the discovery process.

And yet, when one reads historical analyses of a scientific discovery, one is often told with certainty the crucial factors and steps that contributed to the discovery. Such certainties exist, however, only along the particular historical path by which the specific discovery was made. One cannot exclude the possibility that if some of the crucial factors or steps were absent, the discovery might still have eventually occurred. Can, then, such a retrospective analysis be used to investigate a discovery process that did not take place? This is precisely what the question 'Why did modern science not first develop in China?' encourages one to do.

If one examines the steps involved in a scientific discovery in history, one finds that the steps which led to the discovery were often not the logical steps that one would expect based purely on an analysis of the internal consistency of the discovery process. The historical steps, though they may not be random, were certainly not pre-programmed by logic. This implies that the path defined by such historical steps cannot in general be unique in the sense that it is the only path by which the discovery could have been made. The possibility always exists that there are other paths that may eventually lead to the same discovery. Historians can undoubtedly trace the historical path and identify the important steps that led the development of modern science but this provides only an exposition of its development. Certainly, no claim can be made that it is the only path by which modern science could have developed. The multi-path possibility for the development of science makes the analysis of the question of the absence of modern scientific development in a civilization even more elusive.

One might equally well ask 'Why did modern science not first develop in Greece?' Some historians of science might think that this question is not at all parallel to the same question concerning China since they consider the scientific revolution that took place in Europe to be a part of a continuous lineage of development from the Greek civilization. There is no question that the early Greek work in sciences contributed to laying the foundation upon which the scientific revolution began. However, Greek civilization was but one of several civilizations that contributed to this foundation. The scientific revolution and the development of modern science are European achievements.

Consider, for example, the development of the oldest of the physical sciences, namely mechanics, the study of the motion of objects, in the 17th century by Galileo and Newton. These scholars had to first break away from the Aristotelian conceptualization of motion which includes the fallacious concept of 'natural place' implying that an object can exhibit either 'natural' motion in seeking its 'natural place' or 'violent' motion compelled by some external force and the theory of 'antiperstasis' envisioning the motion of an object as being propelled by the air rushing into the 'horror vacui' created by its passage.

In addition, the mathematics that was needed for the mathematization of the physical concepts and hypotheses to provide a quantitative analysis of motion had to be invented by Newton (b. 1642, d. 1727) and Leibniz (b. 1646, d. 1716). And the development of calculus was certainly not in the Greek mathematical traditions nor was the calculus formulated in an axiomatic deductive system. In fact, the rigorous logical foundation of calculus was not laid until the concept of limit was crystallized in the 19th century.

The Chinese contribution to science and technology, unlike that of Greek civilization, has not yet been thoroughly analyzed. There are still many books and historical records in Chinese that have not yet been adequately studied. It will require much more scholarly, interdisciplinary, research before the Chinese contribution to science and technology can be fully documented and understood. It is hoped that this book will initiate further research in the field of science and technology in Chinese civilization.

The San Diego workshop and the publication of this work were funded in part by the University of California at San Diego and the Wei-Kung 爲公 Institute. Additional support was provided by the Program in Chinese Studies of the University of California at San Diego. The willingness of these organizations to support this work is deeply appreciated and gratefully acknowledged. I should also like to express gratitude to Richard E. Attiyeh, Daniel K. S. Chang 張家信, Matthew Y. C. Chen 陳淵泉, Yuan-Cheng Fung 馮元禎, James P. Lin 林伯堅, Shao-Chi Lin 林紹基, and Richard P. Madsen for their interest and assistance in this project. Finally, my thanks must go to Annetta Whiteman for the preparation of the final manuscripts for publication.

Cheng-Yih Chen
程貞一
San Diego, 1986

References

1. Joseph Needham, The Grand Titration: Science and Society in East and West (University of Toronto Press, Toronto, 1969).

2. Nathan Sivin, 'Why the Scientific Revolution Did Not Take Place in China - Or Didn't It?' in Exploration in the History of Science and Technology in China (Shanghai Chinese Classics Publishing House, Shanghai, 1982) edited by Li Guo-Hao 李國豪 et al.

3. A. C. Graham, 'China, Europe and the Origin of Modern Science: Needham's The Grand Titration' in Chinese Science (The MIT Press, Cambridge, Massachusetts, 1973).

MATHEMATICS

A COMPARATIVE STUDY OF EARLY CHINESE AND GREEK WORK ON THE CONCEPT OF LIMIT

Cheng-Yih Chen 程貞一

I. Introduction

The concept of limit is of fundamental importance to modern science. It was the crystallization of this concept in the 19th century, initiated by the works of Bolzano (b. 1781, d. 1848)[1] and Cauchy (b. 1789, d. 1857)[2] that provided the logical foundation for the calculus invented in the 17th century by Newton (b. 1642, d. 1727)[3] and Leibniz (b. 1646, d. 1716).[4] The process for the development of the limit concept began, however, very early in antiquity. The study of this process would reveal not only the historical steps that led to the clarity of the concept but also the fascinating capacities of the human mind, its penetration as well as its vulnerability, as one examines the great minds that had pondered and debated on this concept.

The conceptualization of limit concept was not at all straightforward. It took more than two millenniums and underwent several stages of intuitive understanding before the limit concept was precisely defined by the works of Weierstrass (b. 1815, d. 1897) and Heine (b. 1821, d. 1881).[5,6] In this paper, I examine some of the early work in the Chinese and Greek civilizations on concepts such as continuity, infinite divisibility and the infinitely small which contributed to our understanding of the concept of limit and led to some of the early applications of the limit concepts in mathematics.

In this work, I will make no use of the convention of dating early mathematics books such as the Elements in Greek civilization or the Chiu-Chang Suan-Shu《九章算術》(Nine Chapters on Mathematics) in Chinese civilization to a specific date, since such books underwent several stages of reconstruction and modification. Hence a date of this kind can be rather misleading. Instead, I will, whenever possible, date the content of such books in accordance with its origin. A brief summary of the dates and events of three early mathematics books cited in this work is given in Appendix I.

II. Early Greek Work on the Principle of Infinite Divisibility

Concepts such as continuity, discreteness and infinite processes were among the topics that attracted the interest not only of early mathematicians but also of early philosophers. These concepts were both philosophical issues and mathematical problems. In Greek civilization, the word 'number' meant positive whole numbers. It represented discrete objects. A commensurable ratio therefore represented a relation between two collections of discrete objects. However, the discovery of incommensurable ratios (irrationals) by the Pythagoreans in the -5th century[7] not only brought into focus the difficulties with their philosophical view that all things in the universe depend upon whole numbers but also exposed the logical defects in their theory of proportion. This led to an intense debate on the relation of the discrete to the continuous among Greek philosophers and mathematicians.

Democritus (b. c. -460, d. c. -370) was of the view that all phenomena were to be explained in terms of infinitely small atoms. It was Zeno of Elea (fl. c. -450) who brought this problem into the center of controversy through the paradoxes that he proposed on motion. These paradoxes survived through the work of Aristotle (b. -384, d. -322) who cited Zeno in order to criticize him.[8] We have

(1) You cannot get to the end of a race course.

In motion one must arrive at the half-way stage before one arrives at the goal.

(2) Achilles will never overtake the tortoise.

He must first reach the place where the tortoise started, and by that time the tortoise will have got the some way ahead.

(3) The arrow in flight is at rest.

For if everything is at rest when it occupies a space equal to itself, and what is in flight always occupies a space equal to itself, it cannot move.

(4) Half the time may be equal to double the time.

If we suppose three lines, one (A) at rest, and the other two (B, C) moving in opposite directions, B will pass in the same time twice the number of points in C that it passes in A.

The context in which Zeno proposed these paradoxes is not quite clear. Aristotle's explanation provides very little information.[9]

These paradoxes nevertheless seem to argue against, on the one hand, the view that space and time are infinitely divisible so that motion is continuous and smooth, and, on the other hand, the view that space and time are made up of small indivisible intervals so that motion is a succession of minute jerks. One of the fallacies in Zeno's arguments lies in the incorrect intuition that the sum of an infinite number of positive quantities is infinitely large, even if each such quantity is extremely small. The impact of the discovery of incommensurables and of Zeno's paradoxes on mathematics in Greek civilization was significant. Not only did it redirect the emphasis of the early Greek mathematics from numbers to geometry,[10] but it also excluded the concept of the infinitesimal from Greek work in mathematics.

In Book V of the Elements (see Appendix IC) we have, on the theory of proportion, Definition 4:

> Magnitudes are said to have a ratio to one another which are capable, when multiplied, of exceeding one another.

This definition excluded the possibility of a ratio between two magnitudes in which one is infinitely small, so that a finite multiple of it does not exceed the other. Similarly, this definition also excludes the possibility of two magnitudes in which one is infinitely large so that no finite multiple of the smaller one will exceed the larger.

The concept of magnitude was introduced by Eudoxus of Cnidos (b. c. -408, d. c. -355) to avoid incommensurables. This was accomplished by not giving numerical values either to magnitudes or to ratios of magnitudes. Consequently, magnitudes, in Greek mathematics, were not numbers but stood for time and geometric entities such as line segments, angles, areas and volumes. Contrary to numbers which were discrete, magnitudes were continuous. According to Archimedes (b. -287, d. -212), it was Eudoxus who first established the lemma that magnitudes can be taken to vary continuously.[11] This lemma is found in Book X of the Elements as Proposition 1:

> Two unequal magnitudes being set out, if from the greater there be subtracted a magnitude greater than its half, and from that which is left a magnitude greater than its half, and if this process is repeated continually, then there will be left some magnitude which will be less than the lesser magnitude set out.

In his attempt to prove this theorem, Euclid (fl. c. -300) made use of Definition 4 of Book V mentioned above. Thus, he assumed that magnitudes have properties as described in Definition 4 instead of proving that they do.[12] It is not clear how Eudoxus established this lemma but in Archimedes' work,[11] the theorem was used as an axiom without proof.

Eudoxus's solution to the problem of incommensurables was of fundamental importance to Greek mathematics. It provided the logical foundation for incommensurable ratios in geometric entities and enabled Greek mathematicians to make great progress in geometry. It did not, however, solve the problem of irrational numbers in arithmetic and algebra. As a result, geometry was, to a large extent, isolated from arithmetic and algebra in Greek mathematics.

III. Early Chinese Work on the Principle of Infinite Divisibility

Concepts such as continuity, discreteness and infinite processes were also among the topics of interest to both philosophers and mathematicians in Chinese civilization. Unlike the Greek mathematicians, the early Chinese had a less restrictive concept of numbers.[13,14] As a consequence, there was no counterpart of the incommensurable crisis in Chinese civilization. The 'ming-t'i' 命題[15] on motion devised by dialecticians such as Hui Shih 惠施 and Kungsun Lung 公孫龍 of the Chan-Kuo 戰國 period (-475 to -221) are often identified by historians as the Chinese counterpart of Zeno's paradoxes. Such a comparison is valid only to a certain extent since their content is substantially different.

The 'ming-t'i' of the dialecticians survived through the work of Taoists who cited them in order to criticize the dialecticians. We have, from Chapter 33 of the Chuang Tzu 《莊子》 (The Book of Master Chuang),[16] the 'ming-t'i' on motion:

飛鳥之景未嘗動也。

The shadow of a flying bird never was in motion.

鏃矢之疾而有不行不止之時。

The rapid motion of a flying arrow consists of moments at which [the arrow] is not in motion and not at rest.

The shadow 'ming-t'i' may appear to be similar to Zeno's arrow paradox, since it considers the shadow, like an arrow in flight, not to be in motion. With our current understanding of the work on motion by Mo Ti 墨翟 (b. c. -480, d. c. -390) and his followers from the -5th to the -3rd century, it is now clear that the shadow 'ming-t'i' does not deal with the problem of continuity as previously thought.[17] It deals with the dynamics of motion. Unlike the flying arrow, the moving shadow can neither possess inertia nor have forces acting directly upon it. Based on the dynamic analysis of motion advanced by the Mohists,[18] the dialecticians could successfully argue that dynamically a shadow cannot engage in motion. It derives its apparent motion from the flying bird.

The arrow 'ming-t'i', on the other hand, deals indeed with the problem of continuity in motion. The early Chinese conceptualization of space and time is best stated in the Mo Ching 《墨經》 (The Mohist Canon)[19]:

宇，彌異所也。

久，彌異時也。

Space is that which extends to different positions.

Time is that which extends to different moments.

These definitions consider space and time to be continuous.[20] It then follows that motion in space and time must also be continuous.

The arrow 'ming-t'i' examines the instantaneous aspects of the motion, i.e., the motion at each instant of time which according to the definition given by the Mohists, has no duration. We now know that, at any instant during its flight, the arrow would simultaneously have an instantaneous position and an instantaneous velocity. The arrow 'ming-t'i' states the observation that, at a given instant, the arrow is located at a specific position so that, at that instant, it is not in motion and yet, in the meantime, the arrow is moving (with a velocity) so that, at that same instant, it is not at rest. This is probably one of the earliest attempts to describe the instantaneous behavior of an object in flight in terms of its position and velocity without the help of the mathematical apparatus for taking limits to the infinitesimal time intervals.

The logical foundation for infinite divisibility lies in the concept of the infinitely small. A 'ming-t'i' that deals specifically with this problem was proposed by Hui Shih 惠施 (b. c. -380, d. c. -305). We have[16]:

至大無外，謂之大一，至小無內，謂之小一。

> The infinitely large has nothing beyond itself, and
> is called 'ta-i' 大一.
> The infinitely small has nothing within itself, and
> is called 'hsiao-i' 小一.

This 'ming-t'i' defines the concepts of infinitely large and infinitely small.

With nothing beyond it, 'ta-i' represents the concept of infinitude and boundlessness. An example conceived by the early Chinese of 'ta-i,' the infinitely large, is the universal space-time continuum.[20] 'Hsiao-i' represents that which is not zero yet has nothing within itself. Thus, 'hsiao-i,' the infinitely small, is practically equivalent to the modern notion of the infinitesimal defined to be that which is arbitrarily small and approaches zero as its limit.

The concepts of continuity and the infinitely small provided the necessary logical foundation for Kungsun Lung 公孫龍 (b. c. -320, d. c. -250) to conceive the principle of infinite divisibility. We have[16]:

一尺之棰，日取其半，萬世不竭。

> If from a stick of one foot, one could take away a half [and thereafter a half from the remainder] every day, after ten thousand generations one would still not be able to exhaust the stick.

This is a statement of the principle of infinite divisibility. In other words, a finite magnitude can be the sum of an infinite number of positive quantities. In current mathematical notation, Kungsun Lung's 'ming-t'i' may be stated as:

$$\frac{1}{2} + \frac{1}{2^2} + \frac{1}{2^3} + \cdots + \frac{1}{2^n} + \cdots \longrightarrow 1 \qquad (1)$$

Here we encounter the first adumberation of the concept of limit.[21]

Chinese work on the principle of infinite divisibility, like the Greek work, is not in the form of a theorem with proof. The Chinese approach is, however, based on both the concept of continuity and the concept of the infinitely small. It is less restrictive and, unlike the Greek work, applies equally well to entities other than magnitude. These differences were of significance in its application to various mathematical problems.

IV. Early Work on the Application of the Concept of Limit

A. Application to Irrational Numbers

An irrational number is defined as a number that cannot be expressed as the ratio of two integer numbers. Here we consider the class of irrational numbers representable in terms of roots. A well-known historical example of such an irrational number is $\sqrt{2}$ which has often been interpreted in the context of geometry as the diagonal of a unit square. The Babylonians not only encountered such irrational numbers long before the Greeks and the Chinese, they also found ways to handle them satisfactorily.

In an old Babylonian tablet from the Yale Collection (No. 7289), one finds a diagram of a square with the magnitude of its side, s = 30, and diagonal, d = 42, 25, 35 (base 60):

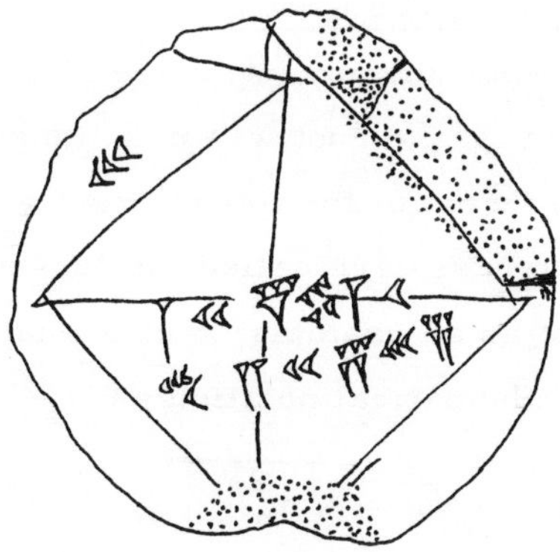

Figure 1. A handcopy of a Babylonian tablet (No. 7289) from the Yale Collection (by Wang T'ien-Luan 王天鸞). The numbers are given in cuneiform symbols in sexagesimal fractions.

The diagram undoubtedly illustrates the relation that the diagonal of a square is $\sqrt{2}$ times its side ($d = \sqrt{2}\, s$) for the third number 1, 24, 51, 10 (base 60) given on the tablet is an excellent approximation for $\sqrt{2}$. Converting this number into base 10, we have:

$$1 + \frac{24}{60} + \frac{51}{3600} + \frac{10}{216000} = 1.414213$$

which is correct to the sixth decimal place.

The early Chinese also appreciated the existence of such irrational numbers but not until the +3rd century before they developed a decimalization procedure to handle them. Instead of sexagesimal fractions, the Chinese made use of decimal fractions. It was in the use of decimal-fraction series to approximate irrational roots that we find an early application of the concept of limit. The use of decimal fractions probably originated from the decimal metrological units which were in use in ancient China before they were standardized in the late -3rd century.[22]

From the Chiu-Chang Suan-Shu《九章算術》(Nine Chapters on Mathematics) (see Appendix IB), it is seen that in early Chinese work on root extraction, if the root was non-integral, the solution would then be given in terms of the integer root with the remainder "left as such."[23] Such cases are called the 'pu-chin' 不盡 cases. Taking the square root as an example, the 'pu-chin' case can be written, in current mathematical notation as

$$\sqrt{x} = \sqrt{a_0^2 + y} \tag{2}$$

where a_0 is the largest integer that can be extracted and y is the remainder. The root would then be given in terms of the integer a_0 and the remainder y. This practice continued until the +3rd century, when Liu Hui 劉徽 proposed, in his commentary on the Chiu-Chang Suan-Shu, a decimalization procedure for root extraction.

Liu Hui remarked that although by keeping the remainder as such the exact value of the root could be maintained the procedure would not be satisfactory. He proposed not to leave the remainder as such but to continue the root extraction in decimal fractions. From page 10b of Chapter 4, we have:

不以面命之，加定法如前，求其微數。

微數無名者，以爲分子，

其一退以十爲母，其再退以百爲母。

退之彌下，其分彌細。

[Let us] not leave [the remainder] as such but continue with the [root-extraction] method as before to obtain additional small numbers [for the root]. Let the 'nameless' small number be the numerator and take 10 to be its denominator if it is to be shifted one position back, and take 100 to be its denominator if it is to be shifted two positions back. The further it shifts back in position, the smaller will be the fraction.

Liu Hui's decimalization for the root may be written, in current mathematical notation, as

$$S_n = a_0 + \frac{a_1}{10} + \frac{a_2}{100} + \cdots + \frac{a_n}{10^n} \tag{3}$$

where n is the number of decimal-fraction terms and the a_i's with $i = 1, 2, \ldots, n$ are the 'nameless' numbers to be shifted i-times back in position. It is obvious that for irrational roots, the procedure could, in principle, be continued indefinitely. Liu Hui further reasoned:

則朱冪雖有所乘之數，不足言之也

Consequently, though there will be some residual values for the 'chu-mi' 朱冪 (i.e. the difference between the square of the root and the square of

> decimal-fraction series, see Eq. (3), of the root), [they can be made so small that] nothing is needed to be said about it.

Liu Hui's remark that "there will be some residual values" indicates that he recognized that the decimal-fraction series could never be made exact since it is only an approximation for the irrational root.

This discussion of the decimalization of roots is based on an intuitive understanding of the concept of limit. Let the irrational root be denoted by R, then Liu Hui's conceptualization of limit may be restated in current mathematic notation as follows:

> For a given value ϵ, the residual difference $R^2 - S_n^2$ can always be made less than ϵ by increasing n

where S_n, defined by Eq. (3), is the series approximation of the irrational root in decimal fractions. This statement of the concept of limit has certain arithmetical features. It is of interest to note that, Cauchy (b. 1789, d. 1857), in his analysis of limit and continuity, considered an irrational number to be "the limit of diverse fractions which furnish closer and closer approximate values of it."[24] In order to be logically consistent, the irrational number should be rigorously defined, independent of the limit process for the irrational numbers.[25]

B. <u>Applications to Problems in Geometry</u>

Let us consider a circle C with an area A and inscribe within the circle a regular n-sided polygon of area P_n. If the number of sides should be doubled, it is obvious that the intermediate area outside the polygon but inside the circle would decrease. It then follows intuitively that by increasing the number of sides successively, the intermediate area can be made arbitrarily small. In current mathematical notation, this intuitive concept of limit can be written as:

$$\lim_{n \to \infty} P_n = A \tag{4}$$

This is the essence of a method that historians of mathematics call the method of 'exhaustion.' In this section, I examine early applications of this method to geometrical problems and their relation to the limit concept in both the Chinese and the Greek civilizations.

a. Early Greek Work

In Greek civilization, the method of 'exhaustion' was ascribed to Eudoxus (b. c. -408, d. c. -355) based on Archimedes' statement that Eudoxus provided the earliest satisfactory proof that the volume of the cone is 1/3 the volume of the cylinder having the same base and altitude. Since, in Greek mathematics, the method of 'exhaustion' was the only method that could provide such a proof, the method is credited to Eudoxus.[26] The earliest written account of the method is found in Book XII of Elements where it is used to prove a number of important theorems.

To illustrate how this method is used in Greek mathematics, let us outline the proof that if A_1 and A_2 are the areas of two circles having diameters d_1 and d_2, respectively, then

$$\frac{A_1}{A_2} = \frac{d_1^2}{d_2^2} \tag{5}$$

To proceed with the proof, first let P_{n1} and P_{n2} be the n-sided polygons inscribed within these two circles, respectively. Then, from Proposition 1 of Book XII,

$$\frac{P_{n1}}{P_{n2}} = \frac{d_1^2}{d_2^2} \tag{6}$$

Now, suppose that we have

$$\frac{A_1}{A_2} > \frac{d_1^2}{d_2^2} \tag{7}$$

We know that the difference between A_1 and P_{n1} can be made so small that the inequality would be maintained if A_1 were replaced by P_{n1}.

$$\frac{P_{n1}}{A_2} > \frac{d_1^2}{d_2^2} \tag{8}$$

With the help of Eq. (6), the inequality given by Eq. (8) can be re-written:

$$\frac{P_{n1}}{A_2} > \frac{P_{n1}}{P_{n2}} \Rightarrow P_{n2} > A_2$$

This is an absurdity since the area of a regular polygon cannot exceed the area of its circumcircle. Hence, the supposition given by Eq. (7) cannot be valid. Similarly, we can show that we cannot have

$$\frac{A_1}{A_2} < \frac{d_1^2}{d_2^2} \tag{9}$$

Consequently, by this method of double 'reductio ad absurdum,' the theorem that areas of circles are to each other as squares on their diameters is established.

The most impressive application of the method of 'reductio ad absurdum' was made by Archimedes (b.c. -287, d.c. -212) to prove that the area A of a parabolic segment BECFD (Figure 2) is

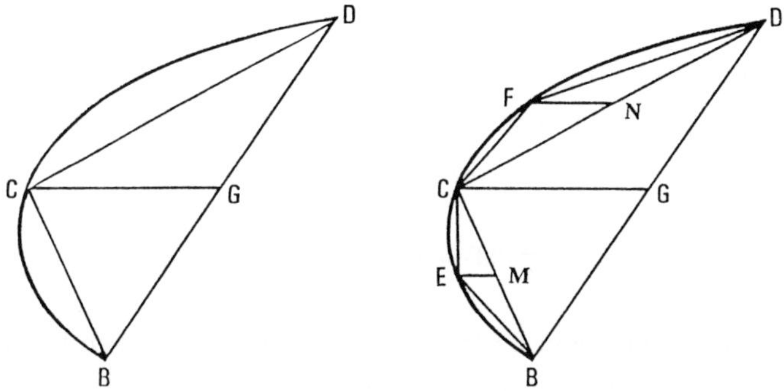

Figure 2. Segment of a parabola.

four-thirds of the area P of an inscribed triangle BCD having the same base and equal height.[27] Let CG be parallel to the axis of the parabola so that it is the diameter bisecting all chords parallel to the base BD of the parabolic segment. Now take BC and CD as the new bases and construct on them triangles in their corresponding arc segments BEC and CFD with EM and FN being diameters parallel to the axis of the parabola. From the geometry of the parabola, Archimedes showed that

$$\Delta BEC + \Delta CFD = \frac{1}{4}\Delta BCD$$

The process of constructing triangles can be repeated by taking BE, EC, CF and FD as the new bases. The sum of the areas of the four triangles so obtained would be $\frac{1}{16}\Delta BCD$. By repeated application of this process, one would obtain the series approximation

$$S_n = \Delta BCD + \frac{1}{4}\Delta BCD + \frac{1}{16}\Delta BCD + \frac{1}{64}\Delta BCD + \cdots$$

In accordance with the limit concept, the area of the parabolic segment can be visualized as the limit of the series since the difference between them can be made arbitrarily small.[28] Archimedes certainly realized that a finite number of terms can approximate the area of the parabolic segment as close as he needed. He, however, avoided the limit process by the use of the method of double 'reductio ad absurdum' to prove his theorem. He showed that there are contradictions in either of the following two statements:

$$A > \frac{4}{3}\Delta BCD \ , \qquad A < \frac{4}{3}\Delta BCD$$

Consequently, the area of the parabolic segment can be neither more nor less than $\frac{4}{3}\Delta BCD$. It then follows that:

$$A = \frac{4}{3}\Delta BCD = \frac{4}{3}P \ .$$

The proof by the method of double 'reductio ad absurdum' is rigorous. But the method is sterile.[29] It requires one to know in advance the relation to be proved. Thus, the method furnishes an elegant tool to establish a known relation but does not lend itself to the initial discovery of the relation. In the above case, for example, the formula $A = \frac{4}{3}P$ had to be determined in advance. Not until the discovery of a copy of Archimedes' lost treatise Method in 1906,[30] did we learn about the method that he used to determine the formula. The method is now known as the method of equilibrium. It was also by this method that Archimedes found the volume of a sphere with radius r to be $\frac{4}{3}\pi r^3$. He then again proved it by the method of double 'reductio ad absurdum.'

It should be noted that 'method of exhaustion' was not the name used by the early Greeks. The term 'exhaustion' was introduced in Europe in the 17th century to depict the fact that the successive inscribed polygons 'exhaust' the area of their circumcircle. The term connotes the concept of infinite processes and the limit concept which were avoided by most Greek mathematicians. It is in Archimedes' work that we find numerical use of the method of 'exhaustion.' He used this method to obtain a higher-order approximation for the value of π, the ratio of the circumference to the diameter of a circle.[31]

To obtain a value for π, Archimedes made use of the proposition that the area of a circle is equal to the area of a right-angled triangle whose base is as long as the circumference of the circle and whose altitude is equal to the radius. He then approximated the circumference more and more closely by the perimeters of the inscribed and circumscribed regular polygons obtained by doubling successively the number of sides. The result for π that he obtained by using polygons of ninety-six sides is:

$$\frac{223}{71} < \pi < \frac{22}{7}$$

This is a remarkable achievement since Archimedes, in carrying out the calculation, had to operate with only whole numbers and fractions.

b. Early Chinese Work[32]

In Chinese civilization, the first indication of a possible use of the method of 'exhaustion' is found in the work of Shang Kao 商高 of early Chou 周. His work was incorporated in the Chou-Pei Suan-Ching 《周髀算經》 (Mathematics Classics of Chou Gnomons) (see Appendix IA) as a dialogue with Chou Kung 周公, the Duke of Chou (regent c. -1026 to c. -1018). We have:

數之法出於圓方。

圓出於方，方出於矩。

矩出於九九八十一。

The mathematical method is derived from [the properties of] circles and squares. A circle is derived from a square and a square from a right-angled ruler ('chü' 矩). A right-angled

> ruler is derived from [the arithmetic principle of] $9 \times 9 = 81$.

In the latter part of the dialogue, Shang Kao again stated:

> 方數爲典，以方出圓。
>
> Taking the mathematics of a square as the guide ('tien' 典), a circle can be derived from a square.

Although these statements are explicit, they provide no information on how a circle was derived from a square.

The method of 'exhaustion' in Chinese civilization is called 'ko-yüan shu' 割圓術 which literally means the method of dissection of a circle. In this method, the circle is visualized as being dissected into pie sectors by a polygon inscribed within it. The pie sectors are then approximated by the corresponding triangular sectors of the polygon. In Liu Hui's 劉徽 work of around +263, we find an explicit use of this method in Chinese civilization.[33] He began with a six-sided polygon inscribed within a circle having a diameter of 2 ft, and showed that the old value $\pi = 3$ would be obtained if one approximates the area of the circle A by P_{12}, the area of a twelve-sided polygon inscribed within the circle. However, to obtain a value of higher accuracy, the dissection process has to be continued by successively doubling the number of sides of the inscribed polygon. Liu Hui remarked[33]:

> 割之彌細，所失彌少。
>
> 割之又割，以至於不可割。
>
> 則與圓周合體而無所失矣。

> The smaller is the size of dissection, the less loss [in area] would be (i.e. the area between the circle and inscribed polygon). If the dissection procedure is repeated to its exhaustion, then [the area of the polygon] would be superimposed on the entire area of the circle so that there would be no loss in area.

This is a clear statement of the limit concept within the context of the method of 'exhaustion.'

After this general discussion of the method of 'exhaustion' and the demonstration that the old value $\pi = 3$ is only an initial approximation of this method, Liu Hui carried out a sample calculation for a higher approximation of π. He obtained the areas of the inscribed polygons with 96 and 192 sides:

$$\begin{aligned} P_{96} &= 3.139344 \text{ ft}^2 \\ P_{192} &= 3.141024 \text{ ft}^2 \end{aligned} \qquad (10)$$

By making the approximation $A \cong P_{192}$, he obtained $\pi = 3.141024$. He then commented that, for ordinary purposes, the value 3.14 (i.e. $\frac{157}{50}$) is a good-enough approximation.

Liu Hui also pointed out that the value for the area A of a circle is always bounded by the inequalities[33]:

$$P_{2n} < A < [P_{2n} + (P_{2n} - P_n)], \quad n = 6, 12, 24, \ldots \qquad (11)$$

This can be seen from Figure 3 since $P_{2n} - P_n = n\frac{1}{2}(BP)(AC)$ and $\frac{1}{2}(BP)(AC)$ is equal to the sum of the triangle areas ΔABQ and ΔCBR. Taking the results given by Eq. (10) as an example, he concluded that the value of π is bounded by the inequalities

$$3.141024 < \pi < 3.142704$$

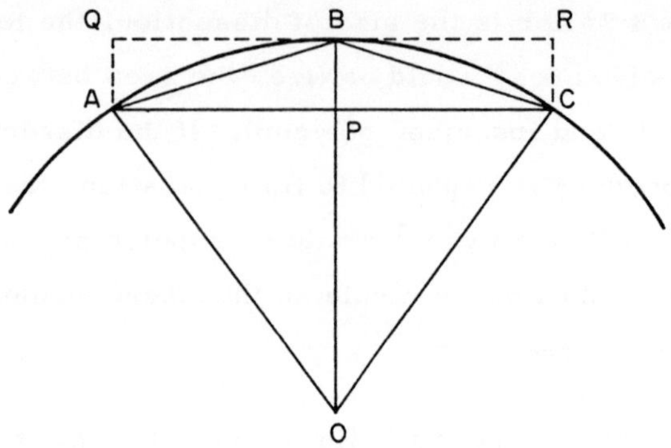

Figure 3. A segment of a circle with a radius OB and sectors of an inscribed polygon.

He further demonstrated this by the determination of a value of even higher accuracy for π and obtained π = 3.1416, correct to the fourth decimal place. As expected, it lies within the bounds. He emphasized that this value is still approximate and inferior to the value obtained by considering a polygon of 3072 sides.[34]

The calculation of the value of π reached its high point in the work of Tsu Ch'ung-Chih 祖冲之 (b. 429, d. 500). Tsu devised a more efficient method, known as the 'keng k'ai' 更開 method.[35] to carry out the 'exhaustion' procedure and obtained the following upper and lower bounds for the value of π.

$$3.1415926 < \pi < 3.1415927 \qquad (13)$$

These values are correct to the sixth decimal place. He further obtained for the value of π, the fractions $\frac{22}{7}$ and $\frac{355}{113}$. These fractions are the convergents of a continued-fraction approximation for π.[36] The first fraction $\frac{22}{7}$, though less accurate, is of historical interest since it was first obtained around -240 by Archimedes. The second fraction $\frac{355}{133}$, known as Tsu's fraction for π, is again correct to the sixth decimal place.

In addition to this numerical application of the method of 'exhaustion,' Liu Hui also made use of the method to improve an approximate formula for the area of a circular segment that appeared in the Chiu-Chang Suan-Shu. Consider a segment of a circle AEDFCB (see Figure 4) with its base chord AC, 'hsüan' 弦, denoted by b_0 and its sagitta height DB, 'shih' 矢, denoted by a_0. The radius of the circle is OD. In Chapter 1 of Chiu-Chang Suan Shu, the area S of the circular segment is given by the formula[37]

$$S \cong \frac{1}{2}(a_0 b_0 + a_0^2)$$

Liu Hui in +263 pointed out that the old formula provides only an approximation and showed that by the method of 'exhaustion' the area of the circular segment can be determined to an arbitrary degree of accuracy.

From the geometry of the circle and of the series of triangles constructed in the exhaustion process (see Figure 4), Liu Hui derived a series for the area of the circular segment[38] (in current mathematical notation):

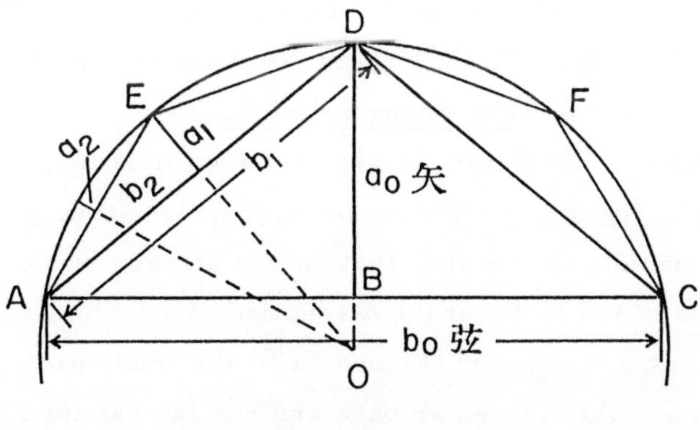

Figure 4. A segment of a circle with an inscribed polygon. OD is the radius of the circle.

$$S = \frac{1}{2}\left(a_0 b_0 + 2a_1 b_1 + 2^2 a_2 b_2 + \cdots 2^n a_n b_n + \cdots\right) \qquad (14)$$

with

$$a_i = \frac{d}{2}\left(1 - \sqrt{1 - \left(\frac{b_i}{d}\right)^2}\right), \quad i = 1, 2, 3, \ldots$$

$$b_i = \sqrt{a_{i-1}^2 + \left(\frac{b_{i-1}}{2}\right)^2}$$

where the a_i and b_i are the sagittas and chords of the series of circular segments generated in the exhaustion process. He reasoned that "if the dissection procedure is repeated until the size of the dissection reaches its extremity in smallness, 'chi hsi' 極細; then the sum of the products of the sagittas and chords would yield a very accurate value [for the area of the circular segment]."

Liu Hui also made use of the limit concept for the determination of the volume of a solid, 'yang-ma' 陽馬, a pyramid with a rectangular base and one lateral edge perpendicular to the base (Figure 5a). The formula for the volume V of a 'yang-ma' given in the <u>Chiu-Chang Suan-Shu</u> calls for the multiplication of the width a, 'kuang' 廣, by the length b, 'mao' 袤, and the height h, 'kao' 高, and the division of the product by three.[39] That is, $V = abh/3$. No proof is given for the formula. The proof was later supplied by Liu Hui in his commentary of the <u>Chiu-Chang Suan-Shu</u>.

In his proof, Liu Hui made use of the formula for the volume U of a 'ch'ien-tu' 壍堵, a right prism with a right-triangular base (Figure 5b), namely $U = \frac{1}{2}abh$, that he had already proven.[40] He also made use of the fact that the dissection of a 'ch'ien-tu' slantwise (Figure 5) gives a 'yang-ma' (Figure 5a) and a 'pieh-nao' 鱉臑, a pyramid with a right-triangular base and one lateral edge perpendicular to the base (Figure 5c). Hence,

$$U = \frac{1}{2}\,abh = V + W \tag{15}$$

where V and W are the volumes of the 'yang-ma' ABCDE and the 'pieh-nao' AFED, respectively.

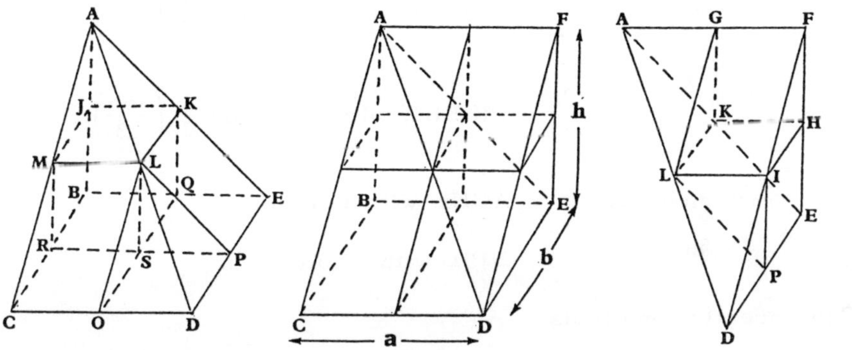

(a) 'Yang-ma' 陽馬 (b) 'Ch'ien-tu' 塹堵 (c) 'Pieh-nao' 鼈臑

Figure 5. (a) Yang-ma, a pyramid with a rectangular base and one lateral edge perpendicular to the base.

(b) Ch'ien-tu, a right prism with a right-triangular base.

(c) Pieh-nao, a pyramid with a right-triangular base and one lateral edge perpendicular to the base.

Following Liu Hui's method of dissection,[41] one obtains for the 'yang-ma' (Figure 5a) the components[42]

'fang ch'i' JKLMRBQS, with a volume $\frac{abh}{2^3}$

'ch'ien-tu' LSQKEP, with a volume $\frac{1}{2}\left(\frac{abh}{2^3}\right)$

LSRMCO, with a volume $\frac{1}{2}\left(\frac{abh}{2^3}\right)$

'yang-ma' AJKLM with a volume V_1

LSPDO with a volume V_1

and for the 'pieh-nao' (Figure 5c) the components

'ch'ien-tu' GKHFIL with a volume $\frac{1}{2}\left(\frac{abh}{2^3}\right)$

PIHEKL with a volume $\frac{1}{2}\left(\frac{abh}{2^3}\right)$

'pieh-nao' AGKL with a volume W_1

DPIL with a volume W_1

Consequently, one finds

$$V = 2\left(\frac{abh}{2^3}\right) + 2V_1 , \qquad W = \frac{abh}{2^3} + 2W_1 \tag{16}$$

If the dissection procedure is repeated for the 'yang-ma' with volume V_1 and the 'pieh-nao' with volume W_1, one finds

$$V_1 = 2\left(\frac{abh}{2^6}\right) + 2V_2 , \qquad W_1 = \frac{abh}{2^6} + 2W_2$$

The procedure can be carried out repeatedly. In general, one obtains

$$V_{n-1} = 2\left(\frac{abh}{2^{3n}}\right) + 2V_n , \qquad W_{n-1} = \frac{abh}{2^{3n}} + 2W_n \tag{17}$$

Hence, the volumes V and W take the form:

$$V = \frac{abh}{2^2} + \frac{abh}{2^4} + \cdots + \frac{abh}{2^{2n}} + \cdots \tag{18a}$$

$$W = \frac{abh}{2^3} + \frac{abh}{2^5} + \cdots + \frac{abh}{2^{2n+1}} + \cdots \tag{18b}$$

Liu Hui reasoned:

半之彌少，其餘彌細。至細曰微，微則無形。

由是言之，安取餘哉？

數而求窮之者，謂以情推，不用籌算。

> The smaller they are dissected, the smaller the remainders. The extremity in smallness is called 'wei' 微 (the minute) which has a vanishing form. From what we have just said, how could the remainders be summed up (ch'ü 取, literally 'taken')? In determining the limiting values [of the volumes], one does not perform the calculations by the use of counting rods but by relying on one's deductive reasoning ('ch'ing t'ui' 情推).[43]

This is a remarkable statement on the limit concept.

When the remainders were properly taken into consideration Liu Hui obtained the formulas:

$$V = \frac{1}{3}\,abh\ , \qquad W = \frac{1}{6}\,abh\ , \tag{19}$$

These results agree with those of the special $a = b = h$ case that he previously considered. The volumes of the 'yang-ma' and 'pieh-nao' given by Eq. (19) can be obtained by taking the limit of the sum given in Eq. (18):

$$V = \frac{1}{4}\,abh\left(1 + \frac{1}{2^2} + \cdots + \frac{1}{2^{2n-2}} + \cdots\right)$$

$$= \frac{1}{4}\,abh\left(\frac{4}{3}\right) = \frac{1}{3}\,abh \tag{20a}$$

$$W = \frac{1}{8}\,abh\left(\frac{4}{3}\right) = \frac{1}{6}\,abh \tag{20b}$$

These same results can also be obtained by deducing from Eq. (18) the theorem that the volume of a 'yang-ma' is twice the volume of a 'pieh-nao.' With this theorem

$$V = 2W \tag{21}$$

These volume formulas can also be obtained from Eq. (15). In his commentary Liu Hui mentioned both the theorem given by Eq. (21) and also the summation of the remainder by deductive reasoning. If Liu Hui did not carry out the summation directly, he certainly realized that the volumes given by Eq. (19) and Eq. (18) are equivalent.

The method of Liu Hui's proof is logically based. Though not an axiomatic deductive proof, it is a direct proof unlike the method of double 'reductio ad absurdum' of Greek mathematics which is an indirect proof. The method of Liu Hui's proof is now known as the method of dissection. The dissection proof in Chinese civilization began with Shang Kao 商高 (see Appendix II) and was called the method of 'chi-chü' 積矩 (lterally, piling up rectangles).[44] Shang Kao used this method to first prove the 'kou-ku' 勾股 theorem, known in the West as the Pythagoras theorem. The dissection proof is appealing to one's intuition and can be readily demonstrated with the aid of models. Liu Hui, in fact, suggested the use of three-dimensional models, 'ch'i' 棊, to assist one's visualization and to verify the result. He also made use of color schemes to identify the models.

Appendix I. Historical Description of Early Mathematics Books

A. The Chou-Pei Suan-Ching 《周髀算經》

The Chou-Pei Suan-Ching 《周髀算經》(Mathematics Classics of Chou Gnomons) is a book in two volumes containing works from three periods. It contains the work of Shang Kao 商高 of the early Chou 周, the work of Ch'en Tzu 陳子 of the Ch'un-Ch'iu 春秋 period, and, towards the end of volume I as well as in volume II, the work of the later 'Kai-T'ien' 蓋天 astronomical school from the Chan-Kuo 戰國 period to the Western Han 漢.

Very little about Shang Kao's personal life is known. Since Shang Kao's work is preserved in the form of a dialogue with Chou Kung 周公 (the Duke of Chou), he probably was a contemporary of Chou Kung in the late -11th century. In the dialogue, Chou Kung addresses Shang Kao by the title 'Ta-fu' 大夫, probably a ranked scholar in the Chou court. The dialogue, composed of a total of 258 Chinese characters, is written in an archaic form, much more so than Ch'en Tzu's work.

Regarding Ch'en Tzu's personal life, again, very little is known. Some information may, however, be deduced from his work. Ch'en Tzu's work is also written in the form of a dialogue, but in this case with his student Jung Fang 榮方. In this dialogue, one finds the following instructions on methodology:

> 夫道術所以難通者。
>
> 既學矣，患其不博。
>
> 既博矣，患其不習。

既習矣，患其不能知。

故同術相學，同事相觀。

The reason why the principle of the method is difficult to master thoroughly is the following:

Although one has learned, one should be concerned about not being broad.

Although one is being broad, one should be concerned about not being able to practice.

Although one is able to practice, one should be concerned about not being able to infer.

Therefore, one learns from similar and related methods, and observes similar and related phenomena.

This methodology is similar to that of the Ju Chia 儒家 , the School of Confucius. In fact, we have from Tsêng Shan 曾参 , one of Confucius's leading disciples, almost the identical passage but in a different context. From the Tsêng Tzu《曾子》(The Book of Master Tsêng) we have the passage:

君子既學子，患其不博也。

既博之，患其不習也。

既習之，患其不知也。

既知之，患其不能行也。

既能行之，患其不能讓也。

Although a 'chün-tzu' 君子 (a man of good character, high attainments and moral excellence) has learned, he should be concerned about not being broad.

Although he is being broad, he should be concerned about not being able to practice.

Although he is able to practice, he should be concerned about not being able to infer.

Although he is able to infer, he should be concerned about not being able to implement.

Although he is able to implement, he should be concerned about not being unpresumptuous.

There is no doubt that these two passages are related. In Tsêng Shen's passage, the methodology is generalized and integrated into the model for the scholarship of a 'chün tzu.' This implies that there was probably a connection between Ch'en Tzu and the School of Confucius. Ch'en Tzu was probably a contemporary of Tsêng Shen in the -5th century, if not earlier

Near the end of Ch'en Tzu's work in the first volume of the Chou-Pei Suan-Ching, it is stated in the text that:

此數望之從周，故曰周髀，髀者表也。

The mathematical art originated from the Chou [capital], it is therefore called Chou Pei 周髀 . Pei 髀 means 'gnomon.'

This statement is consistent with Ch'en Tzu's work since it deals with the estimation of the sun's diameter and the distance between the sun and the earth based on measurements of the sun-shadow length with the Chou 'gnomon' and the mathematics of Shang Kao of early Chou.

Towards the end of Volume I, there are indications of intercalation of the text. This part, together with Volume II, were probably incorporated into the book much later and contains the work of the later part of the 'Kai-Tien' 蓋天 astronomical school. It also contains material similar in content to that of the Lü-Shih Ch'un-Ch'iu 《呂氏春秋》(Master Lü's Spring-Autumn Annals) of -239.

The term 'Chou Pei' 周髀 was also mentioned by Ts'ai Yung 蔡邕 (b. +133, d. +79). The first commentary on the Chou Pei, which is extant, was given by Chao Chün-Ch'ing 趙君卿 of the +3rd century. Subsequently a number of commentaries on this book were written by later mathematicians. The book was selected for inclusion in the Suan-Ching Shih-Shu 《算經十書》(The Ten Mathematics Textbooks) of the T'ang 唐 dynasty (618 to 907) for use by the Mathematics Department of the 'Kuo Chih Chien' 國之監, the Imperial University. The Suan-Ching Shih-Shu were again adopted as textbooks during the Sung 宋 dynasty (960 to 1279) and were officially engraved, printed and distributed to libraries throughout the country in 1084.

B. The Chiu-Chang Suan-Shu 《九章算術》

The Chiu-Chang Suan-Shu 《九章算術》(Nine Chapters on Mathematics) is traditionally identified with the Chan-Kuo period (-475 to -221) of the Chou dynasty. However, the final composition date for the book is probably in the Western Han dynasty (-206 to +8). There have also been proposals placing the date in the Eastern Han (+25 to +220). The discovery in 1984 of mathematical works among the 'chu-chien' 竹簡 ('bamboo slips') unearthed from tombs of the early Western Han permitted the first direct comparison of the Chiu-Chang Suan-Shu with mathematical works preserved directly from the early Western Han.[45] It appears that the final composition date of the book cannot be as late as the Eastern Han.

According to Cheng Chung 鄭衆 (fl. +50) and Liu Hui 劉徽 (fl. +263), the Chiu-Chang Suan-Shu evolved from the Chiu Shu 《九數》(Nine Mathematics) of the early Chou. There is, however, no extant manuscript of the Chiu Shu. The Chiu-Chang Suan-Shu that we have now was reconstructed by Chang Ts'ang 張蒼 (fl. -202 to -162, d. -152) based on fragments of the book surviving from the Ch'in 秦 dynasty (-221 to -207). The book was later edited by Kêng Shou-Ch'ang 耿壽昌 (fl. -75 to -49). The first commentary on the book was given by Liu Hung 劉洪 around +180 but it has long been lost. The second commentary on the Chiu-Chang Suan-Shu, written by Liu Hui 劉徽 in +263, is extant. Subsequently, additional commentaries on the book were written by a number of mathematicians. The book was also selected for inclusion in the Suan-Ching Shih-Shu 《算經十書》 (The Ten Mathematics Textbooks) of the T'ang dynasty.

C. The Elements

The Elements is Euclid's most famous work. Very little about Euclid's personal life is known. According to Proclus (b. +412, d. +485), Euclid lived in Alexandria around -300 and commentaries on his Elements were written by Heron (fl. c. -100), Porphyry (+3rd century) and Pappus (end of the +3rd century). Euclid's work is customarily identified with the Greek classical period. There are no extant manuscripts written by Euclid himself. The Elements we have now was reconstructed in several stages. The early Latin source came from translations by Athelhand of Bath in 1120, Gherard of Cremmona (b. 1114, d. 1187) and Johannes Campanus in 1270 of 8th-century Arabic translations of the Elements. The Campanus translation was the first to be put into print in 1482 in Venice by Erhard Ratdolt, with the second edition (1486) in Ulm and another edition (1491) in Vicenza. In 1505, Bartolemeo Zamberti published in Vienna the first Latin translation directly from Greek based on the +4th-century Theon's recension of Euclid's Elements and in 1533 Simon Grynaeus published at Basel the Greek text, editio princeps. A Latin translation based on the editio princeps was later made by Commandinus in 1572.

In 1808, Francois Peyrand discovered in the Vatican Library a 10th century Greek manuscript (MS Vatican 190). Historians such as J. L. Heiberg considered this 10th century manuscript to be a copy of an edition of Euclid's Elements that preceded Theon's recension and to be a better one since the Elements was probably mostly spoiled by interpolations in about the +3rd century. The Elements was again reconstructed in the 19th century by Peyrand and by Heiberg following the 10th century Vatican manuscript, the best of the Theonine manuscripts and external sources such as the commentaries of Proclus. We have now the Euclidis Opera Omnia (8 volumes and a supplement, 1883-1916) edited by J. L. Heiberg and H. Menge and The Thirteen Books of Euclid's Elements by T. L. Heath (3 vol., 1908, 2nd ed. 1926).

Appendix II. The Dissection Proof of Shang Kao 商高

The mathematical work of Shang Kao 商高 is found in the first section of volume I of the Chou-Pei Suan-Ching (see Appendix IA). The importance of Shang Kao's work and its impact on mathematics in Chinese civilization have not been fully recognized or properly documented. The current debate on the date of the final composition of the Chou-Pei Suan-Ching overshadows the fact that Shang Kao's work was written long before the date of final composition of the book. There is no evidence to indicate any possible alteration or recension of the original 258-word text of Shang Kao's work,[46] or to warrant a change in the dating of this work. Unless there is evidence to deny it, the date for Shang Kao's work remains as that of the Chou Kung 周公 period in the late -11th century, independent of the outcome of the debate.

The most important aspect of Shang Kao's work lies in its use of a proof. He called his method of proof 'chi-chü' 積矩 (literally, 'the piling up of the rectangles'). It is a form of proof now known as a dissection proof. The theorem that he proved is the 'kou-ku' 勾股 theorem, known in the West as the Pythagoras (or Pythagorean) theorem. It is a theorem that summarizes into an analytical mathematical expression, the relation between vertical, horizontal and their hypotenuse. His proof is reproduced below:

故折矩，以爲勾廣三，股修四，經隅五。

既方其外，半之一矩環而共盤。

得成三四五，兩矩共長二十有五，是謂積矩。

> (a) Thus, let us cut a rectangle [diagonally]. [If we] suppose that the width ('kou' 勾) is 3 [units] wide and the length ('ku' 股) is 4 [units] long, then the hypotenuse ('ching' 經) would be 5 [units] long. (b) Now, after drawing a square on the outside [of the hypotenuse], circumscribe it by half-rectangles so as to form a square plate. (c) Then the 3-4-5 relation [for the width, length and hypotenuse] can be established, since the total difference ('kung chang' 共長) between [the square plate] and the two rectangles is an area of 25. This [method of proof] is called 'chi-chü' 積矩 (literally 'piling up the rectangles').

Apparently the first portion (a) of this passage describes the construction of a right-angled triangle from a rectangle and states the 'kou-ku' theorem using its technical terms and a special example. The second portion (b) describes the construction of a diagram for the visualization of the proof given in the last portion (c) of the passage.

In Figure 6, the diagram for the proof of the theorem is reconstructed following the description given in the above passage. A rectangle ABDC is divided into two half-rectangles by hypotenuse CB By drawing a square on the outside of the hypotenuse, we obtain BEFC. A square plate AHIG is then obtained by circumscribing the square BEFC by half-rectangles of the type constructed above. Thus, the hypotenuse diagram that was constructed by Shang Kao for the proof of the 'kou-ku' theorem but which is absent from the text is reproduced.

Following the description given in (c), we examine the difference between the area of the square AHIG, i.e. $(AH)^2$, and the areas

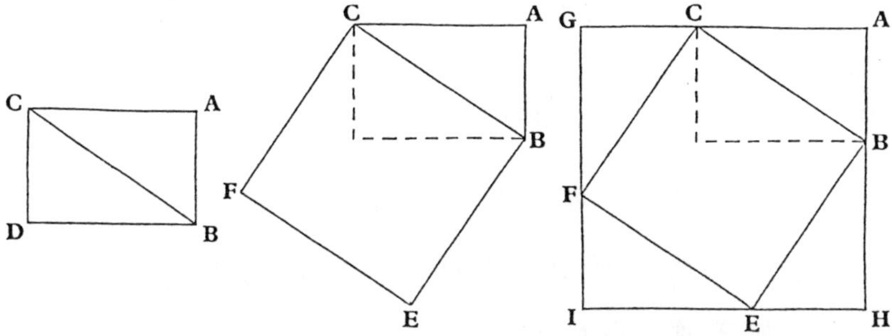

Figure 6. Diagrams for the proof of the 'kou-ku' 勾股 theorem reconstructed according to the prescription given by Shang Kao 商高 in the Chou-Pei Suan-Ching 《周髀算經》 (Mathematics Classics of Chou Gnomons).

of the two rectangles given in terms of the four half-rectangles ABC, BHE, EIF, FGC, i.e. 2(AB)(AC). We find:

$$(BC)^2 = (AH)^2 - 2(AB)(AC) .$$

Since AH = AB + BH and BH = AC, we obtain

$$(BC)^2 = (AB)^2 + (AC)^2 .$$

This is the 'kou-ku' theorem. Thus, we see that the proof of the theorem follows directly from the passage. It does not rely on the numbers given. The specific numbers given in the text were intended as a typical example. This observation is supported by the fact that technical terms for the width, length and hypotenuse are given in the text and by the wording 'i wei' 以爲 ('suppose') as the numbers are introduced as examples for these technical terms.[47]

That Shang Kao's 'chi-chü' dissection proof is a proof for the 'kou-ku' theorem in general was certainly realized by the first commentator on the Chou-Pei Suan-Ching, Chao Shuang 趙爽 (also known under the courtesy name Chün-Ch'ing 君卿) of the +3rd century. This is evident from his commentary as he reconstructed the hypotenuse diagram (see Figure 7) and provided an alternative proof. We have for Chao's proof the passage

> 案弦圖，又可以勾股相乘爲朱實二，
>
> 倍之爲朱實四。以勾股之差自相乘爲中黃實。
>
> 加差實，亦成弦實。

> According to the 'hsüan-t'u' 弦圖 (see Figure 7), one can also let the product of the width ('kou' 勾) and the length ('ku' 股) be two red triangles ($\frac{1}{2}$ ab in Figure 8). The doubling of this gives four

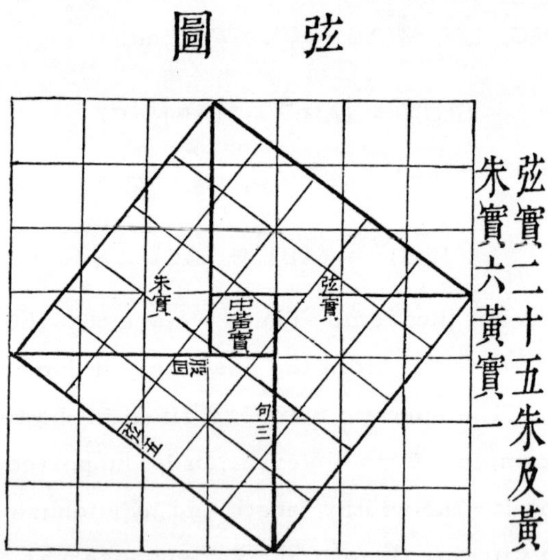

Figure 7. The 'hsüan-t'u' 弦圖 ('hypotenuse diagram') reconstructed by Chao Chün-Ch'ing 趙君卿 for the proof of the 'kou-ku' theorem in his commentary to the Chou-Pei Suan Ching. The areas are labeled by color.

such red triangles. The square of the differences between the length and width gives the center yellow square ($(b-a)^2$ in Figure 8). [Thus], the square of the hypotenuse ('hsüan shih' 弦實) can also obtained by adding together all the areas (i.e. the four red triangles and the center yellow square).

In current mathematical notation, Chao's proof as illustrated in Figure 8 is simply

$$c^2 = 4\left(\frac{1}{2}ab\right) + (b-a)^2 = a^2 + b^2$$

This is a variation of Shang Kao's proof.

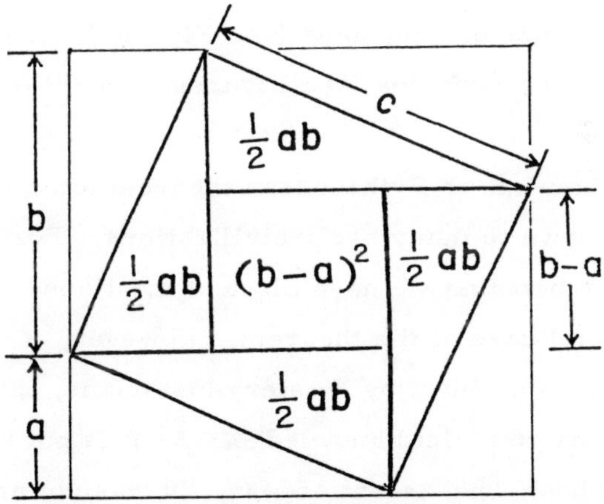

Figure 8. The 'hsüan-t'u' 弦圖 , hypotenuse diagram (see Figure 7) with the color notation replaced by algebraic notation: red triangle = $\frac{1}{2}$ ab and yellow square = $(b-a)^2$.

The interest in Chao's proof lies in its wording, namely "we can also ('yu k'o yi' 又可以) let . . ." and "the square of the hypotenuse can also ('yi' 亦) be obtained. . . ." This indicates that Chao had envisioned his proof as an alternative to Shang Kao's original proof. Therefore, Chao must have fully realized the validity of Shang Kao's proof. But in historical accounts of Chinese work on the 'kou-ku' theorem, Chao's proof is inevitably cited as the proof from the Chou-Pei Suan-Ching. As we have seen, Chao's proof is but a variation of the proof first given by Shang Kao. The same hypotenuse diagram, 'hsüan t'u', later appeared in the work of Bhashara in the 12th century in India.

Shang Kao's 'chi-chü' 積矩 dissection proof is logically based. It makes use of both geometrical and algebraic reasoning. The proof is appealing to one's intuition and can be readily demonstrated with the aid of graphs. To facilitate graphic demonstration, the early Chinese made use of grid lines (see Figure 7) for quantitative visualization and color schemes (see Figures 7 and 8) for geometric visualization.

The 'kou-ku' or Pythagorean theorem for a right-angled triangle was known to many early civilizations. For example, the Babylonian tablet that we have already mentioned (see Figure 1) contains a special case of the theorem. However, it was the proof by Pythagoras, as claimed by Western historians, that raised the theorem above the empirical knowledge. As it is put by Bronowski in his well-known book The Ascent of Man, "It was not until 550 BC or thereabouts that Pythagoras raised this knowledge out of the world of empirical fact into the world of what we should now call proof."[48] If one examines the historical records, the Pythagorean proof is nowhere to be found. However, Bronowski proposed a proof for Pythagoras (see Figure 9). The proof has a strong resemblance to

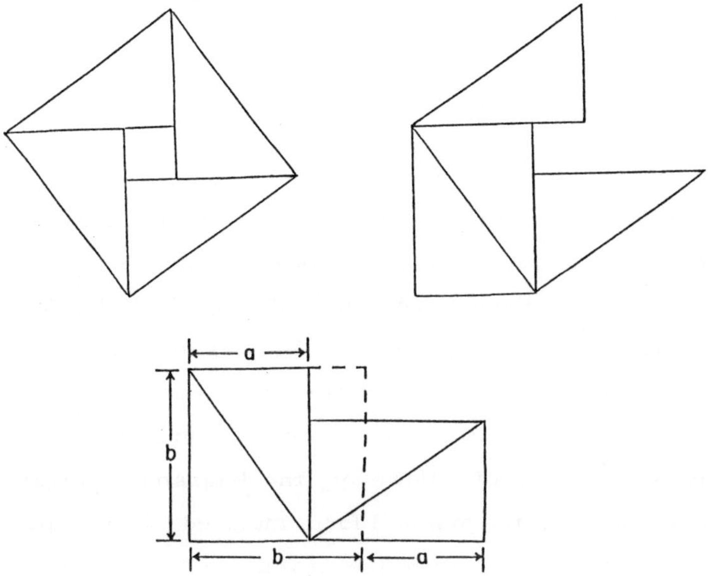

Figure 9. A proposed proof of the Pythagorean theorem by Bronowski (Ref. 48).

the proofs given by Shang Kao and by Chao Chün-Ch'ing, and is, in fact, in the reverse order of a proof first given by Liu Hui 劉徽 in +236 in his commentary to the <u>Chiu-Chang Suan-Shu</u> (see Appendix IB).

We have for Liu Hui's proof the passage[49]:

> 勾自乘爲朱方，股自乘爲青方。
>
> 令出入相補，各從其類。
>
> 因就其餘不移動也，合成弦方之冪。

> The products of the width by itself gives the red square (a^2). The product of the length by itself gives the green square (b^2). Let the difference [between these colored areas and the area of the

hypotenuse square] be patched up by areas of the same shape. This can be done on the remaining portion without moving it. The combined area is that of the hypotenuse square (c^2).

Based on this description of the proof, the diagrams can be reconstructed as shown in Figure 10. It is seen that the diagrams are almost identical to those proposed by Bronowski in 1973. In the late 18th century, an alternative set of diagrams for Liu Hui's proof of the 'kou-ku' theorem was constructed by Li Huang 李潢 (see Figure 11). Either set of diagrams given in Figures 10 or 11 is applicable to Liu Hui's proof. However, the diagrams of Figure 10 are most likely to be in the original form intended by Liu Hui. These diagrams were later known to the Arabic mathematician Tābit ibn Qorra (b. 826, d. 901).[50]

Dissection proofs of this kind did not appear in Europe until very late. The diagram of Figure 10 first appeared in Europe in

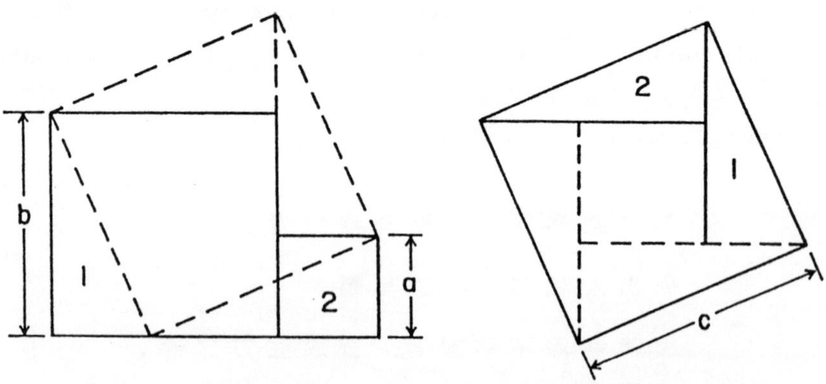

Figure 10. Reconstructed diagrams based on Liu Hui's proof of the 'kou-ku' theorem found in his commentary (+263) to the Chiu-Chang Suan-Shu. (The red square = a^2, green square = b^2).

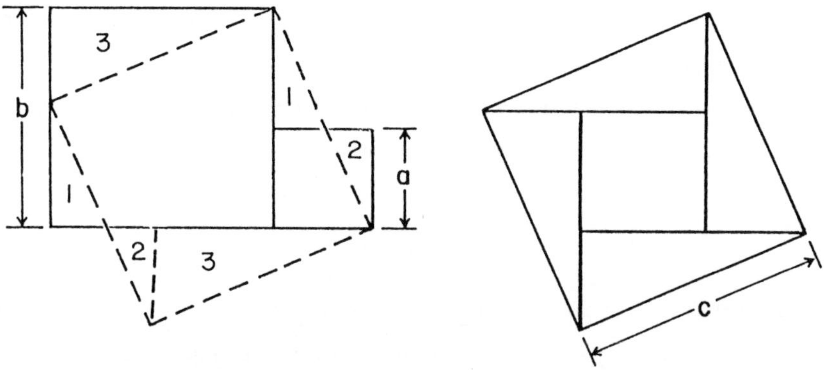

Figure 11. Alternative diagrams constructed by Li Huang 李潢 (d. 1812) for Liu Hui's proof of the 'kou-ku' theorem.

1873 in the work of H. Perigal. Prior to Bronowski, suggestions were already made that such dissection proofs were probably used by Pythagoras or Pythagoreans. However, Heath has stated that this type of dissection proofs is alien to Greek geometrical methods and modes of thinking.[51] The earliest extant proof of the Pythagorean theorem in Greek civilization is that found in the Elements (see Appendix IC). According to Proclus (b. 410, d. 485), the proof found in the Elements is that of Euclid (c. -300). If we scrutinize the claim for Pythagoras's proof of the theorem that bears his name, we find the claims were based primarily on a tradition. The accounts of his proof are colored with mysticism and inconsistency but do not describe the proof.

In Chinese civilization, it has traditionally been claimed that the 'kou-ku' theorem was used in designing projects to control the flooding of the Yellow River in the -21st century by Ta Yü 大禹, Yü the Great, a hydraulic engineer who later became the sovereign-king by establishing the kingdom of Hsia 夏. There is, however, no extant evidence to support this claim. The earliest record of the 'kou-ku' theorem is found in Shang Kao's work. Shang Kao's 高商 'chi-chü' dissection proof of the 'kou-ku' theorem of the -11th century is probably the oldest extant proof of a mathematical theorem. His contribution to mathematics deserves due recognition.

References

1. Bernhard Bolzano, Rein analytischer Beweis des Lehrsatzes, dass zwishen je zwei Werthen, die ein entgegengesetztes Resultat gewähren, wenigstens eine reele Wurzel der Gleichung liege (Gottlieb Hass, Prague, 1817).
2. Augustin-Louis Cauchy, Cours d'analyse algèbrique (1821), in Œuvres (2) (Gauthier-Villars, 1897-99), Vols. III and IV.
3. Isaac Newton, Philosophiae Naturalis Principia Mathematica 1687, translation by Andrew Motte and revised by Florian Cajuri (1934).
4. Gottfried Wilhelm Leibniz, Acta Eruditorum, 3, 467-73 (1684).
5. Karl Weierstrass, Mathematische Werke (Mayer & Müller, Berlin, 1894-1927), 7 vols, and Eduard Heine, 'Die Elemente der Funktionenlehre,' Journal für die Reine und Angewandte Mathematik, 74, 172-188 (1872).
6. For further discussions, see for example, Carl B. Boyer, A History of Mathematics (John Wiley & Sons, New York, 1968).
7. The discovery of incommensurable ratios is attributed to Hippasus of Metapontum of the -5th century. The story was told of how he was later drowned in a shipwreck because he revealed to outsiders of the brotherhood the secret that $\sqrt{2}$ is irrational. The Pythagorean brotherhood was established by Pythagoras (b. c. -585, d. c. -497) and the activity of Pythagoreans spanned the period from c. -550 to c. -400. There are

no written works by the Pythagoreans; we know about them through the writings of later workers.

8. Aristotle, Physics, 239b, 5-33. For the translation of the paradoxes, see for example, J. Burnet in Early Greek Philosophy (Black, London, 1908), p. 367, and in Greek Philosophy (Macmillan, London, 1914), part I, p. 84.

9. Aristotle was of the view that the continuous cannot be made up of the discrete. Hence, he considered that a line is not made up of points since a line is continuous and a point is indivisible even though it has a position on the line. Similarly, he regarded now not being a part of time since now is indivisible and time is continuous.

10. During the Heroic Age of Greek civilization, not only did Pythagoreans place a great emphasis on the role of numbers but mathematicians such as Archytas of Tarentum (b. c. -428) whose students included Eudoxus believed that arithmetic alone could provide satisfactory proofs.

11. Archimedes of Syracuse, Quadrature of the Parabola, -3rd century.

12. For further discussions, see for example, Morris Kline, Mathematical Thought from Ancient to Modern Times (Oxford University Press, New York, 1972), p. 81.

13. The domain of numbers, in the Chinese theory of numbers, included all the positive numbers, 'chêng shu' 正數, and, by the -3rd century, it was extended to include also negative numbers, 'fu shu' 負數. The law of signs for the operation of positive and negative numbers, known as 'chêng-fu shu' 正負術, is given in Chapter 8 of Chiu-Chang Suan-Shu《九章算術》(Nine Chapters on Mathematics) (see Appendix IB). Thus, numbers, as conceptualized by the Chinese mathematicians,

have a continuous domain, extending outward along both the positive and the negative sides of the continuum. The Chinese mathematicians made no formal distinction of irrational numbers from rational numbers and treated them all as numbers.

This mathematical insight of the early Chinese mathematicians has been considered by some historians of mathematics to be the result of logical innocence in failing to appreciate irrationals. But from the early Chinese work on root extraction (-3rd century), it is apparent that they appreciated the existence of irrationals and knew the importance of the remainders in the root extraction for preserving the exact value of the irrationals. In Liu Hui's work (+263) on the decimalization procedure for irrational roots, he stated explicitly that there would always be a residual difference between the root and the series of decimal fractions generated by his procedure to approximate the root, but he reasoned that the value of the residual difference can be made arbitrarily small.

14. For a comparison of early Chinese numeral systems with numeral systems of other ancient civilizations, see Chen Cheng-Yih 程貞一, 'The Development of Numeration Principles in Current Positional Numerals' in the Proceedings of the 3rd International Conference on the History of Chinese Sciences, August 24-24, 1984 (Science Publ. Co., Peking, to be published).

15. 'Ming-t'i' are a class of propositions to be affirmed or denied by debate proposed by dialecticians.

16. The Chuang Tzu is a work of Chuang Chou 莊周 (b.c. -369, d.c. -286). The book we have now also contains work of his disciples.

17. For a review of works on the 'ming-t'i' of dialecticians, see Joseph Needham in Sec. 11 of volume 2 of Science and

Civilisation in China (The University Press, Cambridge, 1956).

18. Joseph Cheng-Yih Chen, 'Early Chinese Work on the Physics of Motion' (to be published), and lecture notes for Chinese Studies 170A entitled Scientific Thought and Intellectual Foundation, University of California, San Diego, 1981 (unpublished).

19. The Mo Ching which contains two canon sections and two sections on canon exposition was compiled in the early -4th century. The canon sections contain a number of Mo Tzu's own writings of the -5th century. The book we have now also contains, however, the work of later Mohists down to the -3rd century.

20. In the elucidation of the definitions of space and time found in the 'exposition' section of the Mo Ching, space is visualized to include east, west, center, south and north; and time to include the past, the present, the morning and the evening. It is obvious that the Mohists' exposition of space and time is not in their most general terms since the given four directions subtend only a two-dimensional space and the examples for time do not include the future. These limitations were soon removed by the work of Shih Chiao 尸佼 (fl. -4th century). We have from the Shih Tzu 《尸子》 (The Book of Master Shih)

宇，四方上下。 宙，往古來今。

> Space includes the four cardinal directions as well as the up and down directions. Time includes the past, the future and the present.

This exposition describes unambiguously a boundless, four-dimensional space-time continuum. It was in this context that the character 'yü' 宇 and 'chou' 宙 were joined toegether to form the binome 'yü-chou' 宇宙 which means the universe.

21. The limit concept contained in this 'ming t'i' has long been recognized by historians of mathematics; for a recent account, see for example, Li Yen 李儼 and Tu Shih-Jan 杜石然, Chung-Kuo Ku-Tai Shu-Hsüeh Chien-Shih《中國古代數學簡史》(Concise History of Mathematics in Ancient China) (Commerce Press, Hong Kong, 1976), p. 25.

22. Before the standardization of decimal units during the Ch'in 秦 dynasty, metrological units were not always advanced by power of 10 even though decimal units were common, see for example, Wu Ch'eng-Lo 吳承洛, Chung-Kuo Tu-Liang-Heng Shih《中國度量衡史》(History of Chinese Metrology) (Commerce Press, Shanghai, 1937).

23. See for example, Ling Wang 王鈴 and Joseph Needham, in T'oung Pao《通報》XLIII, 345-401 (1955).

24. Augustin-Louis Cauchy, Cours d'analyse algèbrique (1821), op. cit. (Ref. 2).

25. Georg Cantor, Mathematische Annalus, 21, 545-91 (1883), and Richard Dedekind, Stetigkeit und irrationale Zahlen (1872) and Was sind und was sollen die Zahlen (1888). English translation of these works in Essays on the Theory of Numbers (Dover reprint, 1963).

26. The method of 'exhaustion' in Greek civilization was also attributed to Antiphon of Sophist (fl. c. -430) and to Hippocrates of Chios (fl. c. -430).

27. Archimedes, Quadrature of the Parabola. See T. L. Heath, The Work of Archimedes (Cambridge, 1897).

28. With our current understanding of infinite series, it is apparent that the series can be summarized to give $\frac{4}{3}\Delta BCD$.

29. See for example, Howard Eves, An Introduction to the History of Mathematics (Holt, Rinehart and Winston, New York, 1953), 3rd edition (1969), p. 320.
30. Archimedes, Method, a treatise addressed to Eratosthenes. The text was restored in 1906 from beneath later writings on a parchment of a 10th-century palimpsest.
31. The work is found in Archimedes' Measurement of a Circle.
32. There are a number of recent accounts of early Chinese work on the concept of limit. These accounts are mostly in Chinese. See for example, Bai Shang-Shu 白尚恕 in Chiu-Chang Suan-Shu and Liu Hui《九章算術與劉徽》(Nine Chapters in Mathematics and Liu Hui) (Peking Normal University Press, 1982) edited by Wu Wen-Jun 吳文俊. The present work is based on the author's lecture notes for Chinese Studies 170 entitled History of Mathematics in Chinese Civilization, University of California, San Diego, 1980 (unpublished).
33. See Liu Hui's commentary (+263) to Problem 32 of Chapter 1 entitled 'Fang-T'ien' 方田 of the Chiu-Chang Suan-Shu, pp. 112 to 14b.
34. The value for π obtained by considering a polygon of 3072 sides is 3.14159. Liu Hui did not give the value explicitly but he was correct in asserting that 3.1416 is inferior to the value obtained from a polygon of 3072 sides. Note that 3.1416 is a better approximation than 3.14158 obtained from a polygon of 1536 sides.
35. The name of Tsu's method is found in Chapter 16 of Sui Shu《隋書》(History of the Sui Dynasty) of c. +646 where these values for π, obtained by Tsu Ch'ung-Chih, are recorded. Tsu's 'keng-k'ai' method is believed to be an improved method of 'exhaustion' which permitted Tsu to obtain a higher accuracy

from a polygon with fewer sides than normally needed. Tsu's method undoubtedly was described in his book entitled Chui Shu 《綴術》 but unfortunately the book has long been lost.

36. It is not clear how these fractions were obtained. Since these fractions are the continued-fraction convergents, it has been suggested that Tsu obtained these fractions by the method of continued fractions.

37. This approximate formula also appeared later in the work of Mahavira (c. 850) in India.

38. From Liu Hui's commentary of +263 on Problem 36 of Chapter 1 of Chiu-Chang Suan-Shu, pp. 17b, 18a.

39. From the method of solution given for Problem 15 of Chapter 5 of Chiu-Chang Suan-Shu, p. 9a.

40. The proof was given in his commentary of +263 on Problem 14 of Chapter 5 of Chiu-Chang Suan-Shu, p. 8b.

41. From Liu Hui's commentary (+263) on Problem 15 of Chapter 5 of Chiu-Chang Suan-Shu, pp. 9a and 9b.

42. The dissection procedure has been studied, for example, by Ch'ien Pao-Tsung 錢寶琮 in the Chung-Kuo Shu-Hsüeh-Shih 《中國數學史》(History of Chinese Mathematics) (Science Publ. Co., Peking, 1964). In the work of Donald Blackmore Wagner [Historia Mathematica 6, 164-188 (1979)] one finds the first correct account of Liu Hui's dissection of the 'yang-ma' and 'pieh-nao' in publication. Parts of Liu Hui's procedure for arriving at these dissections was, however, not fully appreciated by Wagner since he considered those passages in which parts of the procedure are given to be an obscure digression.

43. The translation given here is significantly different from that given by Wagner. His translation is given on pages 181 and

182 of Volume 6 of Historia Mathematica, 1979 and is reproduced below:
"The smaller they are halved, the finer ('hsi' 細) are the remaining [dimensions]. The extreme of fineness is called 'subtle' ('wei' 微), that which is subtle is without form ('hsing' 形). When it is explained in this way, why concern oneself with the remainder? Exhausting the calculation is called 'calculating with the essence'; one does not use calculating-rods to calculate it."

44. See p. 2a to p. 42 of Volume I of Chou-Pei Suan-Ching (see Appendix IA).

45. See 'Chiangling Changchia-Shan Han-Chien Kai-Shu' 〈江陵張家山漢簡概述〉 ('A Description of Han Bamboo Slips Unearthed at Changchia-Shan in Chiangling') in Wen Wu 《文物》 (Cultural Relics) 1, 9-15 (1985) and 'Chiangling Changchia-Shan Shan-Tso-Han-Mu Ch'u-T'u Ta-P'i Chu-Chien' 〈江陵張家山三座漢墓出土大批竹簡〉 ('Bamboo Slips Unearthed from Three Han Tombs at Changchia-Shan in Chiangling') in Wen Wu 《文物》 (Cultural Relics) 1, 1-8 (1985)

46. This observation is also supported by the technical term for the hypotenuse found in Shang Kao's work. The old technical term for hypotenuse was 'ching' 經 and later it changed to 'hsüan' 弦 . We find the old term in Shang Kao's work unchanged even though the new term 'hsuan' is used in Ch'en Tzu's work and the rest of the book.

47. The original translation by Wylie did not translate 'i wei' 以爲 correctly. He translated 'i wei' as 'make.' The term 'i wei' implies a subjunctive mood, 'to suppose,' or 'assuming.'

48. Jacob Bronowski, The Ascent of Man (Little Brown, Boston, 1973), pp. 158-160.

49. From Liu Hui commentary of +263 on Chapter 9 of the Chiu-Chang Suan-Shu.

50. Howard Eves, op. cit. (Ref. 28), p. 71.

51. Thomas Heath, A History of Greek Mathematics (Oxford University Press, Oxford, 1921), pp. 95-100.

MATHEMATICAL FORMALISM IN ANCIENT CHINA

Rong-Zhao Mei 梅榮照

One of the crucial processes in the development of 'modern science' is the mathematization. In considering the question of why modern science did not first develop in China rather than in the West, many historians of science place a great emphasis on the difference between the early Chinese and the early Greek approaches to mathematics.[1-4] The Greek approach to mathematics, especially to geometry, was axiomatic and demonstrative. The Chinese approach to mathematics, on the other hand, is often considered to have been intuitive and computational. In this paper, I attempt to show that the Chinese approach to mathematics was not without proofs and formalism.

I. Descriptions Based on Definitions

During the Warring-States period (-480 to -221), there were debates around the determination of proper terms for certain objects. By 'proper term' was meant to give each object an exact definition. A term could be proper only if it matched the reality; an explanation could be proper only if it was expressed in proper terms.

To stage debates on mathematical concepts, logicians such as Hui Shi 惠施 devised a number of paradoxes: we have, for example,

矩不方，規不可以爲圓。

No square can be drawn with a right-angled ruler,
and no circle can be drawn with a compass.

輪不輾地

A wheel does not touch the ground.

南方無窮而有窮

The south is infinite but also finite.

The first paradox deals with the geometric concepts of the square and circle, the second deals with the concept of the tangent and the third deals with the concept of space.

It was the propositions proposed by the Mohists that successfully defined the mathematical concepts.[5] We have from the Mo Jing 《墨經》 (The Mohist Cannon) the definitions for a circle and a square:

圜，一中同長也。

A circle is a figure with one center having the same length [to the circumference].

方，柱隅四讙也。

A rectangle is a figure whose four straight sides all meet at right angles.

For the definition of a tangent, we have

次，無間而不相攖也。

Two figures touching but not coinciding with no interval between the contact points.

For space, we have

有窮，域不容尺。

無窮，[域]莫不容尺。

Finite space: a region whose edge cannot contain a thread.

Infinite space: a region whose edge could contain a thread.

Liu Hui 劉徽 (fl. +263), who inherited and further developed the Mohists' approach to knowledge, provided for mathematics specific definitions of important mathematical concepts in his commentary to the Jiu-Zhang Suan-Shu《九章算術》(Nine Chapters of Mathematics). He defined the ratio of numbers to be that which has the property of permitting the two numbers to simultaneously increase or decrease by the same factor without affecting it. Based on this definition, he pointed out that the numerator and denominator of a fraction and the two sides of an equation are both ratios. With this definition he was able to define a set of linear equations by the requirement that the two sides of each equation must be a ratio, that the number of linear equations must be the same as the number of unknowns, and that there cannot be equations with the same solutions. Thus, a set of linear equations had solutions only when they met the above requirements. Based on these requirements, Liu Hui pointed out that the 13th problem in Chapter 8 of the Jiu-Zhang Suan-Shu had more unknowns than equations and so the solution should be a ratio.

The law of addition and subtraction of positive and negative numbers in the Jiu-Zhang Suan-Shu was an achievement of great significance. In his commentary, Liu Hui defined positive and negative to be 'two opposite concepts of acquiring and losing.' That was to say, positive and negative were relative in the sense that, if one operation was positive, then the other was negative, and vice versa.

Based on this definition, Liu Hui proposed that the positive and negative numbers in an equation could change signs simultaneously and thus elimination by subtraction and elimination by addition might be combined into a single operation. From Liu Hui's point of view, whenever the definition of mathematical concepts conformed with objective reality, then all algorithms and conclusions that conformed with the definitions were rational.[6]

II. Proofs Based on Axioms

In a sense one may say that all proofs are based on axioms since, in proving a formula or a theorem, one must recognize certain basic assumptions that need no proof. Liu Hui's proofs of formulas and theorems in his commentary on the Jiu-Zhang Suan-Shu were no exception. For example, in the proofs of the areas and volumes of straight line figures, he first recognized a rectangle's area to be the product of its length and width, and the volume of a cuboid (a rectangular parallelepiped) to be the product of its length, width, and height, and then simplified more complicated figures to these basic ones, in order to obtain the areas and volumes of complicated figures. Three pyramids with height, width, and length all equal, could make a cube of the same height, width, and length; thus, the volume of a pyramid equaled one third of that of a cube. By the same reasoning, six right tetrahedrons could make a cube, and thus the volume of such a figure was one sixth of that of a cube. Liu Hui also extended this method to a technique to using infinite divisions to obtain the volumes of figures of different length, width, and height (see Section III below).

From Liu Hui's commentary on the Jiu-Zhang Suan-Shu, we know that the proofs of many formulas and theorems were based on two basic theorems: namely the theorem that corresponding sides of similar right triangles are proportional and the 'gou-gu' 勾股

theorem (known in the West as the Pythagorean theorem). But he did not treat these theorems as axioms. In fact, he gave proofs for these theorems. The proof of the theorem of similar right triangles was simply as shown in Figure 1 where the areas of the small rectangles ab' and ba' are equal since the diagonal line bisects the large rectangle into equal parts. For the proof of the 'gou-gu' theorem, Liu

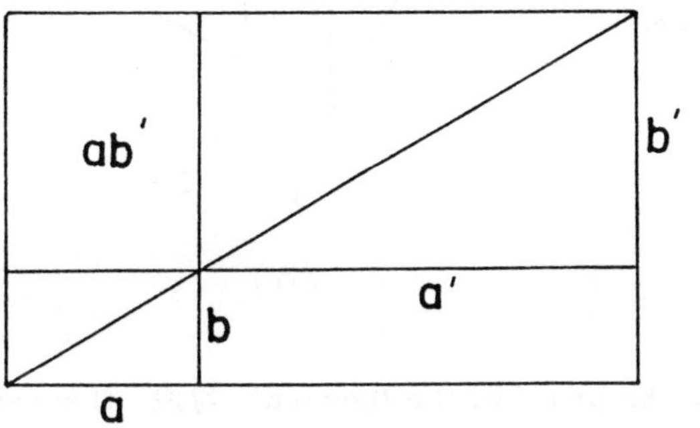

$$ab' = ba'$$

$$\therefore \quad \frac{b}{a} = \frac{b'}{a'}$$

Figure 1. Geometric proof for the theorem of similar right triangles.

Hui provided a different method from that of Zhao Shuang 趙爽 (+3rd century) his contemporary. Zhao Shuang made use of the 'xian tu' 弦圖 (literally, 'hypotenuse graph') given in Figure 2a. But Liu Hui used the method of mutually complementary graphs in Figure 2b. In Liu Hui's method, it is shown that by moving the upper left triangle 1 to the lower right, the middle left triangle 2 to the upper right, and the middle right triangle 3 to the lower left, the area c^2 is obtained from the areas a^2 and b^2.

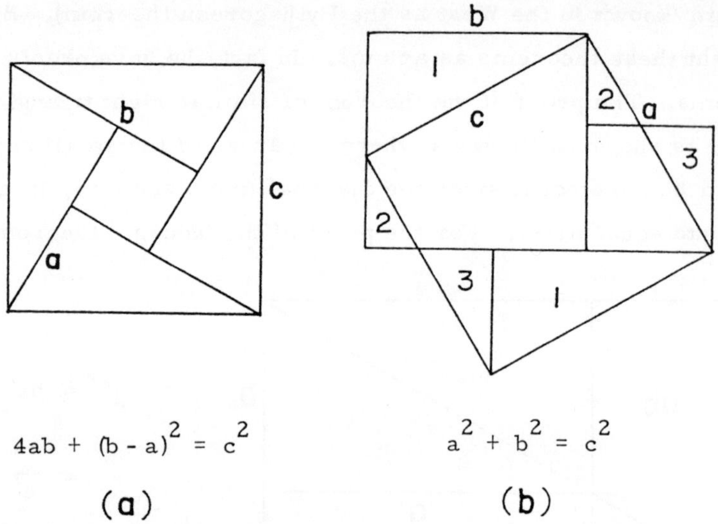

Figure 2. Diagrams for the proof of the 'gou-gu' 勾股 theorem.
(a) The 'xian tu' 弦圖 of Zhao Shuang 趙爽 .
(b) The Liu Hui's 劉徽 method of complementary graphs.

We now consider two explicit examples of Liu Hui's method of proof.

A. The Formula for a Circle Inscribed in a Right-Angle Triangle

The formula for a circle inscribed in a right-angle triangle given in the Jiu-Zhang Suan-Shu is

$$r = \frac{ab}{a + b + c}$$

where r is the radius of the inscribed circle, and a, b, c are the width, height and hypotenuse of the right-angle triangle.

<u>Proof 1</u>. By the construction of a similar triangle with sides a', b', c' as shown in Figure 3, Liu Hui obtained, by similarity

$$\frac{a'}{a} = \frac{a' + b' + c'}{a + b + c}$$

Since

$$a' = r, \quad a' + b' + c' = b$$

Hence

$$r = \frac{ab}{a + b + c}$$

where Liu Hui assumed that $b = a' + b' + c'$ is known.

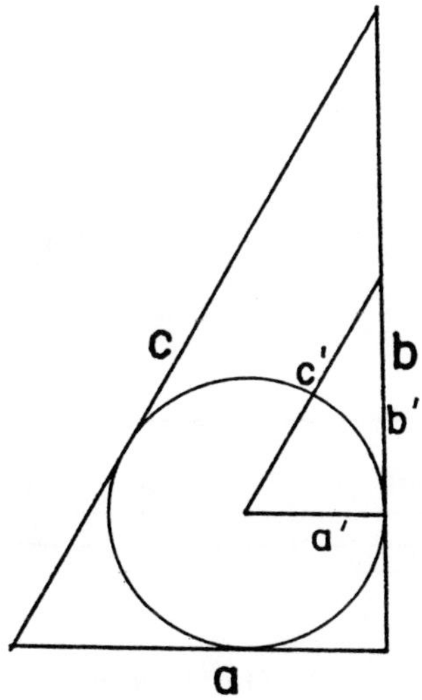

Figure 3. Construction of similar triangles.

<u>Proof 2</u>. Following Liu Hui, it is easily seen that the right-angle triangle with an inscribed circle can be dissected into six pieces of triangles with their widths all equal to the radius of the circle as shown in Figure 4a. By combining each of the six pieces with itself to form a rectangle, one obtains a total of six such rectangles as given in Figure 4b. It then follows by construction that

$$2\left(\frac{1}{2}\, ab\right) = r(c + b + a)$$

Hence

$$r = \frac{ab}{a + b + c}$$

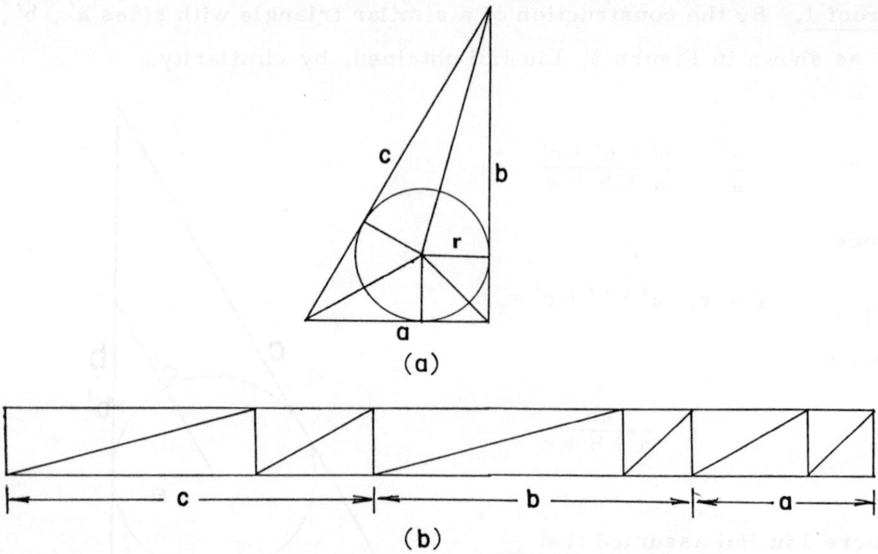

Figure 4 (a) The dissection of a right-angle triangle.

(b) The complementary graph obtained from dissection.

B. <u>The Formulae for 'gou-gu' 勾股 Triples</u>

The formulae for 'gou-gu' triples (known in the West as 'Pythagorean triples') given in the <u>Jiu-Zhang Suan-Shu</u> can be written as

$$a = \frac{n^2 - m^2}{2}, \quad b = nm, \quad c = \frac{n^2 + m^2}{2}$$

where m and n are relatively prime odd numbers, and $a^2 = b^2 = c^2$.

<u>Proof 1</u>. Making use of the relation $b^2 = (c - a)(c + a)$ that he had already proven, Liu Hui obtained

$$c - a = \frac{b^2}{c + a}$$

Let $b = km$, $a + c = kn$ with $n > m$, then $c - a = \frac{km^2}{n}$. Solving these equations, he obtained

$$a = \frac{k(n^2 - m^2)}{2n}, \quad b = km, \quad c = \frac{k(n^2 + m^2)}{2n}$$

The formulae are obtained by setting $k = n$.

Proof 2. Follwoing Liu Hui's description, diagrams for the proof of the formulae for 'gou'gu' triples can be constructed as shown in Figure 5. The diagram provides the algebraic relations

$$2c(c + a) = (c + a)^2 + b^2,$$

and

$$a(c + a) = (c + a)^2 - (c + a)c.$$

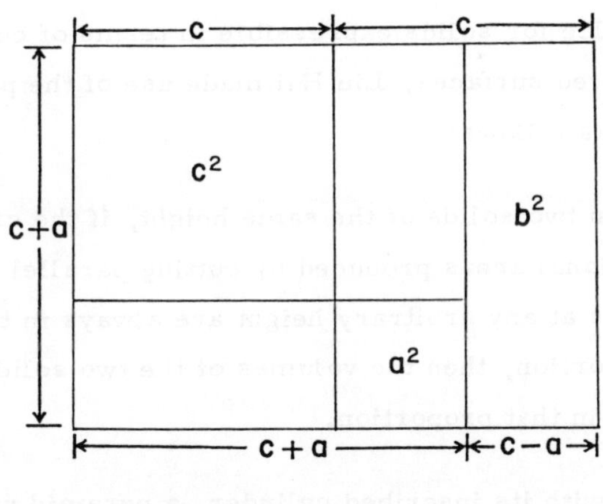

$$b^2 = (c - a)(c + a)$$

Figure 5. Diagram constructed to establish algebraic relations for the proof of formulae for 'gou-ku' 勾股 triples.

It then follows, from these algebraic relations, that

$$c = \frac{(c+a)^2 + b^2}{2(c+a)}, \qquad a = \frac{(c+a)^2 - (c+a)c}{c+a}$$

Again let $b = km$, $a + c = kn$ with $n > m$, then

$$a = \frac{k(n^2 - m^2)}{2n}, \qquad b = km, \qquad c = \frac{k(n^2 + m^2)}{2n}$$

As before, the desired formulae are obtained by setting $k = n$.

From the above examples, it can be seen that Liu Hui used two kinds of proofs. One was derivational, making use of basic theorems and algebraic operations to directly derive the needed formula (proof 1). The other one was graphic, making use of geometric graphs to examine the known formula (proof 2). These were Liu Hui's favorite methods.[7]

The method of complementary graphs was also used by Liu Hui to prove formulae for solids expressible in terms of cuboids. For solids with curved surfaces, Liu Hui made use of the principle which may be stated as follows:

> Given two solids of the same height, if the cross-sectional areas produced by cutting parallel to their bases at any arbitrary height are always in the same proportion, then the volumes of the two solids are also in that proportion.

Since a cuboid with its inscribed cylinder, a pyramid with its inscribed circular-base cone, and a frustum of a pyramid with its inscribed frustum of a cone all have a ratio of $4/\pi$ in their cross sectional areas at an arbitrary height, they must also have a ratio of $4/\pi$ in their volumes. Thus, the formulae for the volumes of a cylinder, a cone with a circular base and a frustum of a cone were easily obtained. Based on this proposition, Liu Hui pointed out that the formula for the volume of a sphere given in the Jiu-Zhang Suan-Shu is incorrect since the ratio $4/\pi$ between cross-sectional areas should be for the ratio between the volumes of a 'mou-he fang-gai' 牟合方蓋 (see Figure 6) and its inscribed sphere instead of the ratio between the volumes of a cylinder and its inscribed sphere. Liu Hui could not, however, determine the volume of the 'mou-he fang-gai.' Based on Liu Hui's proposal, Zhu Chong-Zhi 祖冲之 and his son Zhu Geng 祖暅 obtained the formula for the volume of a sphere V by determining the volume of the 'mou-he fang-gai' as the difference of the volume of its exterior tangential cube D^3 and the total volume of two pyramids, $2\left(\frac{1}{6}D^3\right)$. Hence,

$$V = \frac{\pi}{4}\left[D^3 - 2\left(\frac{1}{6}D^3\right)\right] = \frac{\pi}{6}D^3$$

where D is the diameter of the sphere.[8]

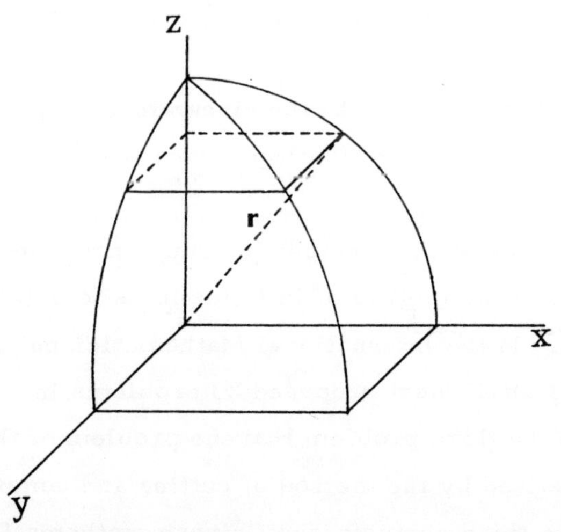

Figure 6. One eighth of a 'mou-he fang-gai' 牟合方蓋 with a horizontal cross sectional area.

III. Proofs Using Limits

The areas of n-sided polygons ($n \geq 3$) can be derived by reducing them to rectangles; but the area of a circle cannot be derived this way. It was known to the ancient Greeks and Chinese that, when the number of sides in a polygon increased infinitely, the figure would become a circle. However, when proving formulas for the area of the circle, mathematicians of these two civilizations made use of different methods. Archimedes used an exhaustion method — supposing the area of a circle to be greater than or less than πr^2, and proving these led to contraditions and thus the area of a circle could only be πr^2. Liu Hui, on the other hand, used a limit method and directly calculated the area of a polygon to be the product of its radius and half its circumference. When the number of sides was increased infinitely, the circumference equalled that of the circle. Thus, the area of the circle was the product of its radius and half its circumference, symbolically,

$$S = \lim_{n \to \infty} S_n = \lim_{n \to \infty} \frac{1}{2} \ell_n r = \frac{1}{2} Lr$$

where S was the area and L the circumference of the circle, S_n was the area of the inscribed polygon, and ℓ_n the circumference of the polygon.

C. F. Gauss of the nineteenth century proposed that it was necessary to use the method of infinitesimals to obtain the volume of a tetrahedron. In the International Mathematicians' Conference in Paris in 1900, D. Hilbert proposed 23 problems in mathematics. He pointed out in the third problem that the problem of the tetrahedron could not be solved by the method of cutting and composing. However, as early as the third century, the Chinese mathematician Liu Hui had already recognized that it was not rigorous to use the method of constructions to check the volumes of the rectangular pyramid and the tetrahedron, he used a method of infinite division to solve the problem. The procedure is as follows:

> A prism with a right triangle base is constructed from a right-triangular tetrahedron and a rectangular pyramid. This prism is divided equally to obtain four small prisms similar to the original (two in the upper rear, and two in the lower front) and two identical rectangular parallelipipeds (which can each be constructed from two small prisms) (Figure 7). Besides the left upper rear and right lower front small prisms, the remaining 3/4 of the total volume contains two rectangular parallelepipeds in the original rectangular pyramid and one rectangular parallelepiped in the original right-triangular tetrahedron which have a ratio of 2:1 in volume. The same method is used to

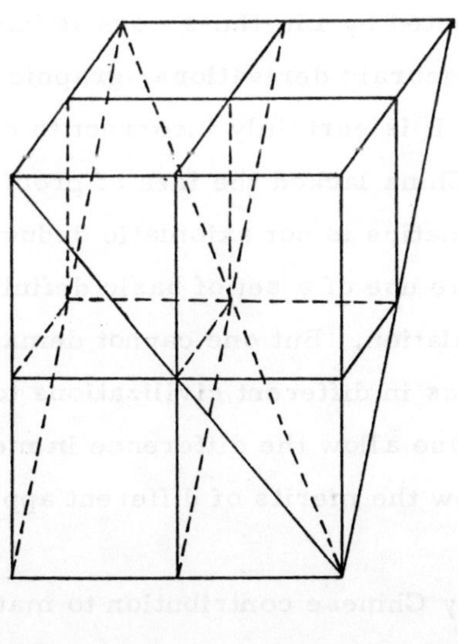

Figure 7. Dissection of a right triangular prism.

> divide the remaining small prisms. It can be shown that in 3/4 of their volumes, the ratio of the volumes of the small rectangular pyramid and the small right-triangular tetrahedron is again 2:1. Four smaller prisms are produced from this step. Repeatedly applying this method, it can be shown that, in the whole prism, the volumes of the pyramid and the tetrahedron are 2:1. Thus, it is proved that the volume of the former is 1/3 abh, and that of the latter is 1/6 abh (a, b, h are the length, width, and height, respectively, of the rectangular parallelepiped).

Liu Hui's method of using infinite division in proofs agrees with Gauss' method of infinitesimals; it also does not contradict Hilbert's proposition that cutting and composing could not be used.[9] In proving the area of a bow given in the Jui-Zhang Suan-Shu, Liu Hui also used a limits method; this work will not be repeated here.

At this point, it is worthwhile to note that the methodology and formalism represented by Liu Hui's work is based on definitions and proofs involving algebraic derivations, graphic verification, and limit procedures. It is certainly incorrect to consider that mathematics in ancient China lacked the idea of proofs. The Chinese formulation of mathematics is not axiomatic deductive since it does not systematically make use of a set of basic definitions and axioms.as in the Greek formulation. But one cannot demand that the development of mathematics in different civilizations to follow the same pattern. Nor should one allow the difference in methodology and formulation to overshadow the merits of different approaches in mathematics.

Another early Chinese contribution to mathematics is indeterminate analysis. In the Sun-Zi Suan-Jing 《孫子算經》 (Master

Sun's Mathematical Manual), we find the first worked-out problem in linear congruences. This work is known in the West as the Chinese remainder theorem. The correctness of this theorem can be easily verified by determining the 'cheng lu' 乘率 ('the multiplication rate'). Given, in modern notation,

$$N \equiv R_i \pmod{a_i},$$

the 'cheng lu' k_i are determined by the relation

$$k_i \frac{M}{a_i} \equiv 1 \pmod{a_i}, \qquad M = a_1, a_2, \ldots, a_n .$$

An early application of linear congruences is in the Chinese calendrical science. Although it is still controversial as to how Zhu Chong-Zhi 祖冲之 extended Liu Hui's approach to the problem of π and derived his famous ratio 355/113, the author thinks that the method Zhu used is connected with linear congruences and theories of fractions.

IV. Mathematics in the Song-Yuan 宋元 Period

The emergence of the Song algebraists brought ancient Chinese mathematics to its highest peak. We now examine their approach to mathematics. Their work on numerical solutions to higher order equations, on higher order arithmetic progressions and on interpolation methods were all developed based on the Jia-Xian 賈憲 triangle of binomial coefficients (known in the West as the 'Pascal triangle'). Jia Xian's original work has been lost, but Yang Hui 楊輝, in quoting this table of coefficients in his book Xiang-Jie Jiu-Zhang Suan-Fa 《詳解九章算法》 (A Detailed Analysis of the Mathematical Method in the 'Nine Chapters') of 1261, named the table 'Kai-Fang Zuo-Fa Ben-Yuan' 開方作法本源 ('The Origin of the Method for Extraction of Roots') with the note

> 出釋鎖算書，賈憲用此術。
>
> The method came from the 'shi suo' 释鎖 (literally 'the key to unlocking binomial coefficients') mathematics book. Jia Xian 賈憲 made use of this method.

and appended the 'Zang Cheng Kai Fang Qiu Lian Cao' 〈增乘開方求廉草〉('Explanation of the Zeng-Cheng Method for Finding the Coefficients). The steps again for finding coefficients to the sixth power are as follows:

1	1 + 5 = 6				
1	1 + 4 = 5	5 + 10 = 15			
1	1 + 3 = 4	4 + 6 = 10	10 + 10 = 20		
1	1 + 2 = 3	3 + 3 = 6	6 + 4 = 10	10 + 5 = 15	
1	1 + 1 = 2	2 + 1 = 3	3 + 1 = 4	4 + 1 = 5	5 + 1 = 6
(1)	1	1	1	1	1

Clearly, this method could be extended to arbitrary order and is a general method for the generation of coefficients. The Jia-Xian triangle as quoted by Yang Hui is reproduced in Figure 8. Later Zhu Shi-Jie 朱世傑 in his Si-Yuan Yu-Jian《四元玉鑑》(Precious Mirror of the Four Elements) of 1303 further stated that the sum of two consecutive numbers in a row is equal to the number in the next row at the intersection of the two lines from those numbers and that in the left leading diagonal, the first k numbers sum up equal to that at the lower right of the kth number (the right leading diagonal is the opposite).[10]

In their approach to higher numerical equations and higher equations with multiple variables, the Song algebraists introduced symbols to represent unknowns. They adopted characters such as 'tian' 天, 'di' 地, 'ren' 人, and 'wu' 物 to denote unknown

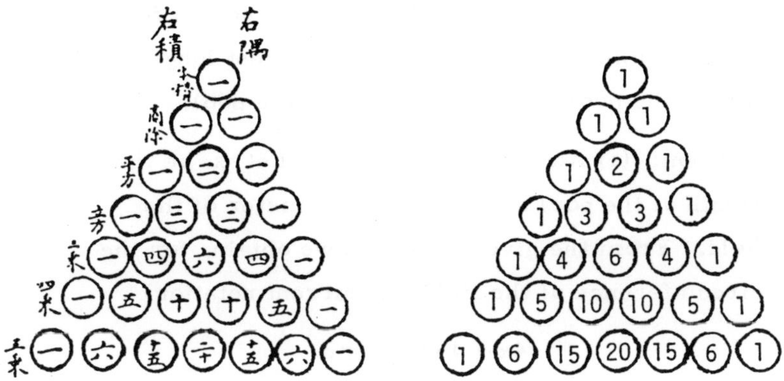

Figure 8. The Jia-Xian 賈憲 triangle of the 11th century as quoted by Yang Hui 楊輝 in his Xiang-Jie Jiu-Zhang Suan-Fa《詳解九章算法》(A Detailed analysis of the Mathematical Methods in the 'Nine Chapters') of 1261.

elements x, y, z and w. This permitted them to establish algebraic relations in terms of unknowns and to demonstrate the solutions by solving these relations for the unknowns. One version of this method is known as the 'tian-yuan shu' 天元術 which has been translated literally as 'the method of celestial elements.' In this method, the coefficients of an equation were called the 'tiao duan' 條段 and the derivation of the 'tiao duan' was called 'yan duan' 演段. The algebraic relations were usually derived by algebraic operations with graphic illustrations.

An example from Li Ye's 李冶 book Yi Gu Yan Duan《益古演段》(The Derivation of the Coefficient for Yi Gu) of 1259 illustrates the method.

> Given: A circular field with a square pool at the center (see Figure 9a), the area outside of the pool

is 8096 square feet (literally 33 arcs and 176 feet), the circumference of the field is 152 feet longer than that of the pool. Find the length of the pool.

Solution: (a) Algebraic derivation: Let x be the length of the pool and d be the diameter of the field. From the given condition we obtain, for the circumference of the field (in current notation)

$$4x + 152\ \text{ft} = \pi d$$

On squaring, one obtains

$$(4x + 152\ \text{ft})^2 = \pi^2 d^2$$

But $\pi^2 d^2 = 4\pi\left(\frac{\pi}{4} d^2\right)$, i.e. just 4π times the total area of the field plus the pool. Thus,

$$(4x + 152\ \text{ft})^2 = 4\pi\,(x^2 + 8096\ \text{ft}^2\,)$$

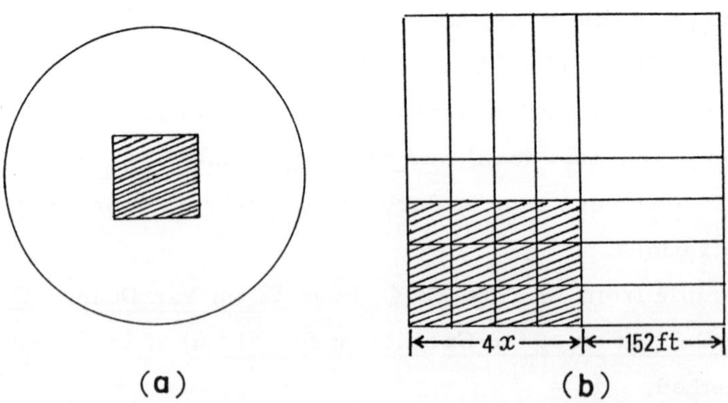

Figure 9. (a) A circular field with a square pool at its center.
(b) Square whose side is equal to the circumference of the circular field of (a).

where $8096\ ft^2$ is the area of the field not including the pool as given. After simple algebraic operation, the equation can be rewritten as

$$4x^2 + 8(152\ ft)x - [12(8096\ ft^2) - (152\ ft)^2] = 0$$

where π is taken to be the nearest integer 3.

(b) Geometric derivation: The area obtained by squaring $4x + 152$ ft is given in Figure 9(b). It is seen that the area has 16 square pools, 8 rectangles with 152 ft as their length and the length of the pool as their width, and a square whose side is 152 feet. Since it is given that $4x + 152$ ft is equal to πd, and $\pi^2 d^2$ is 4π times the total area of the field plus the pool, one obtains

$$4\pi(8096\ ft^2) + 4\pi x^2 = 16x^2 + 8(152\ ft)x + (152\ ft)^2$$

which is the same as that obtained by the algebraic derivation given above.

Based on this discussion, it is clear that the Song-Yuan mathematicians, in total agreement with their traditions, demanded algebraic derivation and geometric verification for mathematic proofs.[11]

It is natural and certainly appropriate that, as the algebraic method reached its full acceptance, geometric verification was gradually eliminated. Consequently, the reliance on algebraic derivation in Li Ye's 李冶 Ce Yuan Hai Jing《测圓海鏡》(Sea Mirror of Circle Measurements) of 1248, Qin Jiu-Shao's 秦九韶 Shu-Shu Jiu-Zhang《數書九章》(Mathematical Treatise in Nine Chapters) of 1247, and Zhu Shi-Jie's 朱世傑 Si-Yuan Yu-Jian《四元玉鑑》(Precious Mirror of the Four Elements) of 1303 cannot be considered to be a lack of proofs. The Song-Yuan mathematicians followed a logical structure in their approach to mathematics. Although the structure is not axiomatic deductive, it is by no means intuitive and lacking in proofs.

V. The Introduction of the Greek Axiomatic System in China

In 1607, the first six volumes of Euclid's Elements were translated into Chinese by the Italian missionary Matteo Ricci and the Chinese 'Da Xue Shi' 大學士 Xu Guang-Qi 徐光啓 under the title Ji-He Yuan-Ben 《幾何原本》(Elements of Geometry). Xu advocated that "everybody in the world should learn about it" and attempted to reformulate traditional Chinese mathematics in the axiomatic system. His representative works are the Ce-Liang Yi Tong 《測量異同》(Similarities and Differences in Measurements) and Gou-Gu Yi 《勾股義》(The Meaning of the Right Triangle). However, this attempt to use the formalism of the Elements to reformulate Chinese mathematics lasted only for a short time.

After the downfall of the Ming 明 dynasty, the Elements was still highly regarded in the early Qing 清 dynasty. Emperor Kang Xi 康熙 even invited Western missionaries to explain the book. Famous mathematicians and astronomers such as Wang Xi-Chan 王錫闡, Mei Wen-Ding 梅文鼎, and Li Shan-Lan 李善蘭 all devoted, at one time or another, research efforts to this book. But the attitudes of the mathematics of the time were much different from that of Xu Guang-Qi as interest in traditional mathematics renewed. Mathematicians of that time were interested in interpreting the Elements within the framework of Chinese mathematics rather than in adopting the axiomatic formulation for Chinese mathematics. In the book Ji-He Tong-Jie 《幾何通解》(A General Methodology in Geometry), for example, Mei Wen-Ding used ancient Chinese methods for solving right-angle triangles to prove 15 theorems in volumes II, III, IV, and VI of the Elements. Although Mei virtually ignored the axiomatic formulation, his work was not without Western influence. This is apparent since he presented his work by first giving definitions, then derivations and finally applications. During the periods of Emperors Qian-Long 乾隆 and Jia-Qing 嘉慶, Jiao Xun 焦循 made use of the associative, cumulative, and distributive laws of addition, subtraction, multiplication and division to reformulate the algorithms in classical mathematics works such as the Jiu-Zhang Suan-Shu. Wang Lai 汪萊 and Li Rui 李銳 studied the problem of positive roots of higher equations and obtained results similar to those of Descartes. There is no doubt that these works were influenced by Western mathematics but none of them were presented in the axiomatic system.

It is worth asking whether the axiomatic formulation played an important role in the development of modern science and mathematics as proposed by some experts on the history of science. The author thinks that this question deserves further discussion. First, let us consider the development of modern mathematics. We have to bypass calculus since the emergence of calculus had nothing to do with the axiomatic system. A number of historians of science consider that analytic geometry was a product of the combination of algebra and geometry. According to Joseph Needham, analytic geometry was developed when Europeans used the algebraic methods developed by the Indians and Chinese in the deductive geometry developed by the Greeks and their successors. He is of the point of view that "this was the greatest single step ever made in the progress of exact science." It is clear, when we examine books by Descartes and Fermat, the founders of analytic geometry, that the geometrical works of Archimedes and Apollonious on conic sections were of importance to the development of analytic geometry. But analytic geometry was not developed axiomatically. Ideas of analytic geometry are also found, without axiomatic considerations, in the work of

Li Shan-Lan entitled 'Jian Zhui Shu' 尖錐術 ('Method of Pointed Pyramids').[12] The process of mathematization was of great importance to the development of modern science. But the axiomatic system is not a prerequisite for the development of the mathematics that permitted this mathematization process.

References

1. Yoshio Mikami, Mathematics in China and Japan (Chelsea Publ. Co., New York, 1913).

2. Joseph Needham, Science and Civilisation in China (Cambridge University Press, 1959), Vo. 3, p. 91.

3. See the review by Nathan Sivin of Needham's Science and Civilisation in China, V. 4, Part 3 in Scientific American, 1, 113 (Jan. 1972).

4. Yabuuchi Kiyoshi, 'The Tradition and Character of Chinese Science,' Science and Philosophy 1, 62 (1984) (translated by Jiang Zhen-Huan) Zi Ran Bian Zheng Fa Tong Xun 《自然辯証法通訊》(Journal of Dialectics of Nature) (1984).

5. Mei Rong-Zhao 梅榮照, 'The Logic in the Mo Jing and Its Influence upon the Ancient Chinese Mathematics,' the XVIIth International Congress of History of Science, University of California, Berkeley, July 31-August 8, 1985.

6. Mei Rong-Zhao 梅榮照, 'Liu Hui de Shu-Xue Li-Lun' 《劉徽的數學理論》 (Liu Hui's Theory of Mathematics') in Zi Ran Bian Zheng Fa Tong Xun 《自然辯証法通訊》(Journal of Dialectics of Nature) 6, 47-48 (1982).

7. Mei Rong-Zhao 梅榮照 'Liu Hui de Gou-Gu Li-Lun' 〈劉徽的勾股理論〉('Liu Hui's Theory of Right Triangle') in Ke-Xue-Shi Ji-Kan 《科學史集刊》(Collected Papers on the History of Science) 11, 77 (1984).

8. Mei Rong-Zhao 梅榮照 'Liu Hui Yu Zhu Chong-Zhi Fu-Zi' 〈劉徽與祖冲之父子〉 ('Liu Hui Versus Zhu Chong-Zhi and His Son') in op. cit. (Ref. 7), 11, 105 (1984).

9. Mei Rong-Zhao 梅榮照, op. cit. (Ref. 6), pp. 50-51.

10. Qian Bao-Zong 錢寶琮, Zhong-Guo Shu-Xue Shi 《中國數學史》 (History of Chinese Mathematics) (Science Publ. House, Beijing, 1964), pp. 150-152.

11. Mei Rong-Zhao 梅榮照, 'Li Ye Ji Qi Shu Xue Zhu Zuo' 冶及其數學著作〉 ('Li Ye and His Mathematical Works') in Song Yuan Shu Xue Shi Lun Wen Ji 《宋元數學史論文集》 (Collected Papers on the History of Mathematics in the Song and Yuan) (Science Publ. House, Beijing, 1966), pp. 145-146.

12. Mei Rong-Zhao 梅榮照 and Wang Yu-Sheng 王渝生, 'Jie Xi Ji He Neng Zai Zhong Guo Chan Sheng Ma?' 《解析幾何能在中國產生嗎?》 ('Could Analytical Geometry Have Been Invented in China?') in Ke Xue Chuan Tong Yu Wen Hua 《科學傳統與文化》 (Scientific Tradition and Civilization) (Science & Technical Publ. House, Shaanxi 陝西, 1983), p. 267.

RESEARCH ON TSU CH'UNG-CHIH'S 祖冲之 APPROXIMATE METHOD FOR π

You-Liang Zha 查有樑

Tsu Ch'ung-Chih 祖冲之 (429-500) was a distinguished mathematician and astronomer in ancient China. His major mathematical work, the Chui Shu 《綴術》, was lost long ago. However, his work on π (the ratio of the circumference of a circle to its diameter) survived its incorporation into the Sui Shu 《隋書》 (History of the Sui Dynasty) by Wei Cheng 魏徵 et al. We have from Chapter 16 of the Sui Shu the following upper and lower bounds for the value of π:

$$3.141592 < \pi < 3.141592 \tag{1}$$

and the following ratios for the value of π

$$\begin{aligned} \pi &\cong \frac{355}{113} \quad \text{'mi-lü' 密率 (accurate ratio)} \\ &\cong \frac{22}{7} \quad \text{'yüeh-lü' 約率 (inaccurate ratio)} \end{aligned} \tag{2}$$

Unfortunately, the account given in the Sui Shu provides no information as to how these values for π were obtained.

Many suggestions have been made attempting to explain how Tsu obtained these values. Among the methods that have been suggested[1,2] are the method of exhaustion, the method of finite differences, and the method of continued fractions. There are also suggestions that the values were obtained, for example, by the use of indeterminate equations or infinite series. No doubt the method was described in his book Chui Shu. According to Shen Kua 沈括 of the 11th century, 'chui shu' 綴術 was a mathematical procedure that had something to do with calendrical science.[3]

In this paper, I re-examined the problem and propose an interpolation procedure for the calculation of the value for π. It is suggested that this procedure probably was Tsu's procedure.

I. The Method of Exhaustion

Ch'ien Pao-Tsung 錢寶琮 assumed that Tsu Ch'ung-Chih used the method of exhaustion in his computation of the values for π.[1] This had been Liu Hui's 劉徽 method for finding π and involves increasing the number of sides of a polygon inscribed within a circle (see Figure 1).[4] Liu Hui showed that the area of a circle S is

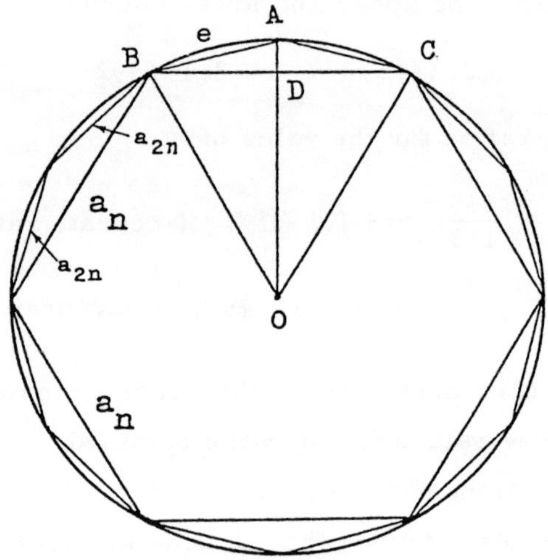

Figure 1. Illustration of the method of exhaustion by the inequalities.

bounded by the inequalities

$$S_{2n} < S < S_{2n} + (S_{2n} - S_n) \qquad (3)$$

where S_{2n} is the area of a regular polygon of 2n sides inscribed within the circle. Liu Hui also pointed out that for a unit circle (i.e.

a circle with a radius of 1 unit) the coefficiency obtained for the circle's area is π [i.e. $S = \pi\,(\text{unit})^2$].

Following Liu Hui's procedure, the length of each side of an inscribed regular polygon of 2n sides, a_{2n}, can be generated from the length of each side of an inscribed polygon of n sides, a_n, using

$$a_{2n} = \sqrt{2r\left[r - \sqrt{r - \left(\frac{a_n}{2}\right)^2}\right]} \qquad (4)$$

where r is the radius of the circle. Taking r = 1 unit, we have

$$a_{2n} = \sqrt{2 - \sqrt{4 - a_n^2}}$$

The area of an inscribed regular polygon of 2n sides is then obtained by the formula

$$S_{2n} = 2n\left[\frac{1}{2}\left(\frac{a_n}{2}\right) r\right] = \frac{1}{2}\, n a_n r \qquad (5)$$

Ch'ien Pao-Tsung found, for n = 12288, the areas of the polygon:

$$S_{12288} = 3.14159251\ (\text{units})^2$$

$$S_{24576} = 3.14159261\ (\text{units})^2$$

Using the inequalities given by Eq. (3), he then obtained Tsu's values for the upper and lower bounds of π given in Eq. (1).[1]

We note that Liu Hui's recursion procedure requires high numerical precision. It would have been difficult for Tsu to apply the recursion formulae twelve times with counting rods while maintaining the level of precision required to obtain the desired number of significant figures. For this reason, a number of historians have conjectured that a modified procedure requiring less recursion with higher accuracy was probably used by Tsu Ch'ung-Chih. The procedure was likely related to the method by which Tsu obtained the famous fraction 355/113 for π.

II. A Proposed Interpolation Procedure

Liu Hui's inequalities given by Eq. (3) can be rewritten as

$$S_{2n} < S < S_{2n} + \Delta S_n$$

where $\Delta S_n = S_{2n} - S_n$ is 2n times the area ΔABD in the nth iteration (see Figure 1). Hence

$$\Delta S_n = 2n(\Delta ABD)_n \tag{6}$$

where $(\Delta ABD)_n$ is the area that corresponds to the area ΔABD in the nth iteration. Geometrically, it is clear that the lower bound of the inequalities could be interpolated upwards adding $2nD_n$ where D_n is a deficit approximation for the arc area AeB in the nth iteration. The interpolated inequalities can be written as

$$S_{2n} + \alpha_n \Delta S_n < S < S_{2n} + \beta_n \Delta S_n \tag{7}$$

with the coefficients α_n and β_n given by

$$\alpha_n = \frac{D_n}{(\Delta ABD)_n} \qquad \beta_n = \frac{E_n}{(\Delta ABD)_n} \tag{8}$$

These coefficients can be determined by continued fractions.

From Liu Hui's work, Tsu knew that, in the 96th iteration, the area of a circle is given by the approximation

$$S \cong S_{192} + \frac{S_{192} - S_{12}}{S_{12} - S_6}(S_{192} - S_{96}) . \tag{9}$$

Rewriting this equation in the form

$$S \cong S_{192} + \gamma_{96} \Delta S_{96} , \tag{10}$$

it follows that

$$\gamma_{96} = \frac{S_{192} - S_{12}}{S_{12} - S_6} \quad . \tag{11}$$

Comparing Eq. (10) with Eq. (7), it is seen that the coefficients α_n and β_n can be obtained by interpolating γ_n.

The values for the polygon areas are

$$S_{192} = 314\,\frac{64}{625}\left(\frac{\text{units}^2}{100}\right) \qquad S_{96} = 313\,\frac{584}{625}\left(\frac{\text{units}^2}{100}\right)$$
$$S_{12} = 300\left(\frac{\text{units}^2}{100}\right) \qquad S_6 = \sqrt{67500}\left(\frac{\text{units}^2}{100}\right) \tag{12}$$

With these values, we obtain

$$S \cong 3.1416 \text{ ft}^2 , \qquad \gamma = 0.3508 .$$

By continued fractions, we obtain for γ

$$\gamma_{96} = \frac{3508}{10000} = \frac{1}{2+}\,\frac{1}{1+}\,\frac{1}{5+}\,\frac{1}{1+}\,\frac{1}{4}\cdots$$

and the asymptotic fractions

$$\frac{1}{2},\ \frac{1}{3},\ \frac{6}{17},\ \frac{7}{20},\ \frac{32}{97},\ \ldots$$

Thus, the approximate values for α_n and β_n can be obtained from the asymptotic fractions

$$\alpha_{96} = \frac{1}{3} \qquad \beta_{96} = \frac{6}{17}$$

since $\alpha_{96} < \gamma_{96}$ and $\beta_{96} > \gamma_{96}$.

If we approximate α_n by α_{96} for $n > 96$, and β_n by β_{96} for $n > 96$, then the interpolated inequalities can be written as[5]

$$S_{2n} + \frac{1}{3}\Delta S_n < S < S_{2n} + \frac{6}{17}\Delta S_n$$

Since $\Delta S_n = S_{2n} - S_n$, we have

$$\frac{4}{3}S_{2n} - \frac{1}{3}S_n < S < \frac{23}{17}S_{2n} - \frac{6}{17}S_n$$

where 4/3, -1/3, 23/17 and -6/17 are coefficients of interpolation. From Tsu's commentary to the Chiu-Chang Suan-Shu 《九章算術》 (Nine Chapters of Mathematics), it is known that he carried out the calculation to $n = 1536$ with

$$S_{1536} = 3.14158389 \text{ units}^2$$

$$S_{3072} = 3.14159046 \text{ units}^2$$

This gives

$$S_{3072} + \frac{1}{3}(S_{3072} - S_{1536}) = 3.14159265 \text{ units}^2$$

$$S_{3072} + \frac{6}{17}(S_{3072} - S_{1536}) = 3.14159278 \text{ units}^2$$

and Tsu's inequality becomes

$$3.1415926 < \pi < 3.1415927$$

The interpolation procedure by continued fractions can also be used to obtain fractions for approximating the value of π. Beginning again with the 96th iteration given by Eqs. (9) and (12), one obtains

$$S \cong \frac{3927}{1250} \text{ units}^2 = 3.1416 \text{ units}^2$$

By continued fractions, we obtain for π

$$\pi \cong \frac{3927}{1250} = 3 + \frac{1}{7} + \frac{1}{16} + \frac{1}{11}$$

and the asymptotic fractions

$$\frac{3}{1}, \frac{22}{7}, \frac{355}{113}$$

These are Tsu's fractions for π. A flowchart of the algorithm for continued fractions and convergents is given in Figure 2.

Based on our study, Tsu's method 'chui shu' 綴術 (which is also the title of his mathematics book) is possibly an interpolation method based on continued fractions and convergents. The method was probably developed in calendrical science. There are indications that a form of interpolation was used by Lohsia Hung 落下閎 in his calculation of the T'ai Ch'u 太初 calendar of -104.[6]

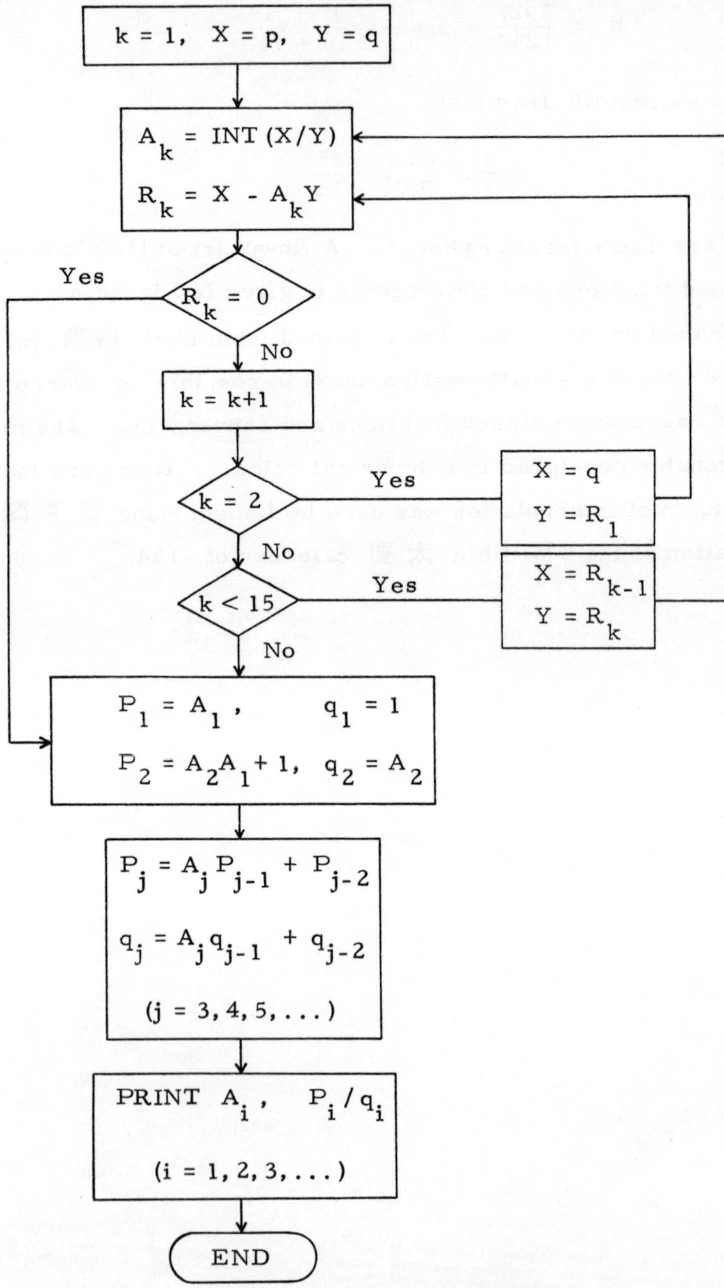

Figure 2. A flowchart for a continued-fraction algorithm.

References

1. Ch'ien Pao-Tsung 錢寶琮, Chung Kuo Shu Hsueh Shih《中國數學史》(History of Chinese Mathematics) (Science Publ., Peking, 1964).

2. Mei Rong-Zhao 梅榮照, 'Liu Hui 劉徽 versus Tsu Ch'ung-Chih 祖冲之 and His Son' in Shu-Hsueh-Shih Lun-Wên Chi《數學史論文集》(Collected Papers on the History of Mathematics) (Peking, 1984), pp. 105-129.

3. Shen Kua 沈括, Meng-Ch'i Pi-T'an《夢溪筆談》(Meng-Ch'i Essays) (Sung 宋 dynasty, 1086).

4. Commentary by Liu Hui 魏徵 (fl. 263) on the Chiu-Chang Suan Shu《九章算術》(Nine Chapters of Mathematics).

5. The same values 1/3 and 6/17 are obtained if we replace S_{192} in Eq. (11) by the better value 3.1416.

6. Zha You-Liang 查有樑, 'A Comparison Between Ptolemy's System and Lohsia Hung's 閣下宏 system.' Presented at the XVIIth International Congress of History of Science, University of California, Berkeley, July 31-August 8, 1985.

WANG LAI'S 汪萊 RESEARCH ON NUMBER SYSTEMS OF VARIABLE BASE

Zhao-Hua Li 李兆華

Wang Lai 汪萊 (1768-1813) was a distinguished Chinese mathematician during the Qing 清 dynasty (1644-1911). He was born in She 歙 county, Anhui 安徽 province. In 1854, his disciple Xia Xie 夏燮 published his miscellany entitled <u>Heng-Zhai Suan-Xue Yi-Shu He-Ke</u>《衡齋算學遺書合刻》(<u>Posthumous Collection of the Mathematical Work of Heng Zhai</u>)[1] which included 7 volumes of his mathematical works as well as papers on music, history, literature, etc.[2]

In this miscellany, there is an important paper entitled 'San-Liang Suan-Jing'〈參兩算經〉('The Mathematics of the Rules of Two and Three'). The paper was written around 1792 and is a study of number systems of variable base, p ($p \leq 10$).[3] In the paper, two main problems are discussed, the first being multiplication and multiplication tables in systems of different bases, and the second being discussion and the definition and calculation of 'proper denominators.'

Wang separates p into two cases: odd numbers and even numbers. He gives multiplication tables for both odd and even numbers. For the odd numbers, for example, $p = 9$, we have

$$8 \times 2 = 17\ , \quad 8 \times 3 = 26\ , \quad 8 \times 4 = 35\ ,$$
$$8 \times 5 = 44\ , \quad 8 \times 6 = 53\ , \quad 8 \times 7 = 62\ ,$$
$$8 \times 8 = 71\ .$$
$$7 \times 2 = 15\ , \quad 7 \times 3 = 23\ , \quad 7 \times 4 = 31,$$
$$7 \times 5 = 38\ , \quad 7 \times 6 = 46\ , \quad 7 \times 7 = 54\ .$$
$$6 \times 2 = 13\ , \quad 6 \times 3 = 20\ , \quad 6 \times 4 = 26\ ,$$
$$6 \times 5 = 33\ , \quad 6 \times 6 = 40\ .$$

$5 \times 2 = 11$, $5 \times 3 = 16$, $5 \times 4 = 22$,

$5 \times 5 = 27$.

$4 \times 2 = 8$, $4 \times 3 = 13$, $4 \times 4 = 17$.

$3 \times 2 = 6$, $3 \times 3 = 10$.

$2 \times 2 = 4$.

For the even numbers, for example, p = 8, we have

$7 \times 2 = 16$, $7 \times 3 = 25$, $7 \times 4 = 34$,

$7 \times 5 = 43$, $7 \times 6 = 52$, $7 \times 7 = 61$.

$6 \times 2 = 14$, $6 \times 3 = 22$, $6 \times 4 = 30$,

$6 \times 5 = 36$, $6 \times 6 = 44$.

$5 \times 2 = 12$, $5 \times 3 = 17$, $5 \times 4 = 24$,

$5 \times 5 = 31$.

$4 \times 2 = 10$, $4 \times 3 = 14$, $4 \times 4 = 20$.

$3 \times 2 = 6$. $3 \times 3 = 11$.

$2 \times 2 = 4$.

As for division in base p systems, Wang mainly discussed denominators. He proposed the concept of a 'proper denominator' defined as follows:

> If a number A is divided by an integral number q in the base p system and $\frac{A}{q}$ is an integral number or finite decimal, then the denominator q is a proper denominator in this base p system.

Wang gave all of the proper denominators for every base p system in which $p \leq 10$ (see Table 1) excluding the obvious proper denominators 1 and p.

Table 1. Proper Denominators for Base p Systems ($p \le 10$)

p	System	Proper Denominators
9	nonary	3
7	septenary	none
5	quinary	none
3	ternary	none
10	denary	$5, 2, 2^2, 2^3$
8	octonary	4, 2
6	senary	$3, 2, 2^2$
4	quaternary	2
2	binary	none

Wang pointed out that the proper denominators for $p \le 10$ were arrived at by dividing by 3 when p was an odd number and by 2 when p was an even number. For example, when $p = 9$, 9 divided by 3 gives 3. Hence, the only proper denominator is 3 in a nonary number system. When $p = 10$, 10 divided by 2 equals 5 while 10 divided by 5 equals 2. Furthermore, $2^2 = 4$ and $2^3 = 8$. Therefore the proper denominators are 5, 2, 4, and 8 in the denary system.

Obviously, Wang's procedure is related to prime decomposition for p. As the fundamental theorem of arithmetic we have, for positive integers $p \ge 1$, the prime decomposition

$$p = p_1^{\alpha_1} p_2^{\alpha_2} \cdots p_n^{\alpha_n}, \quad n \ge 1 \tag{1}$$

where the p_i are prime factors and the α_i are either zero or positive integers. For a positive number q such that

$$q = p_i, \quad i = 1, 2, \ldots, n$$

we know that p/q must be a positive integral number, that is to say, p/p_i is a positive integral number in the base p system. In addition, $1/p_i$ must be a positive finite decimal, otherwise p/p_i would not be a positive integral number. Thus, $(1/p_1)^{\alpha_i \beta_i}$ is equal to 1 or a positive finite decimal with $\alpha_i \beta_i$ equal to zero or a positive integral number. When

$$q = \left(p_1^{\alpha_1}\right)^{\beta_1} \left(p_2^{\alpha_2}\right)^{\beta_2} \cdots \left(p_n^{\alpha_n}\right)^{\beta_n} \tag{2}$$

we get

$$\frac{A}{q} = A \left(\frac{1}{p_1}\right)^{\alpha_1 \beta_1} \left(\frac{1}{p_2}\right)^{\alpha_2 \beta_2} \cdots \left(\frac{1}{p_n}\right)^{\alpha_n \beta_n} \tag{3}$$

Hence A/q will be an integral number or finite decimal if A is an integral number or finite decimal.

If we want to find the proper denominators in a base p system, we can use Eq. (2). Taking different values for α_i and β_i, we can obtain the set of q which are proper denominators. As an example, let us consider the denary number system. We have

$$p = 10 = 2 \times 5$$

$$q = (2)^{\beta_1} (5)^{\beta_2}$$

taking

$$\beta_1 = 0,\ 0,\ 0,\ 1,\ 1,\ 2,\ 2,\ 3,\ 3,\ \ldots$$

$$\beta_2 = 0,\ 1,\ 2,\ 0,\ 1,\ 0,\ 1,\ 0,\ 1,\ \ldots$$

then

$$q = 1,\ 5,\ 25,\ 2,\ 10,\ 4,\ 20,\ 8,\ 40,\ \ldots$$

By selecting $1 < q < 10$, we have

$$q = 5,\ 2,\ 4,\ 8\ .$$

It should be noted that the condition $p \leq 10$ specified by Wang is the necessary condition for the validity of his procedure. Obviously, if one considers a number, say $p = 35$, that violates the condition $p \leq 10$, the prime decomposition given by Eq. (1) cannot be obtained by Wang's procedure of dividing the number by 3. Besides, Wang only took $1 < q < p$, if one takes for q a number from Eq. (2) that is larger than p then q is also a proper denominator in this base p system, which is not included in Wang's work.

Ever since antiquity numeral systems with different bases have been known. The Mesopotamian numeral system, for example, was sexagesimal (base 60) while the Egyptian numeral system was denary (decimal). The Mayan numeral system, on the other hand, was a vigesimal (base 20) system superimposed on a quinary (base 5) system. The study of the mathematical properties associated with different bases is, however, rare. Prior to the development of electronic computers in the 1940's, only a few mathematicians such as G. W. Leibniz (1646-1716) had explored this subject. Wang's research is of significance in the history of mathematics. The concept of a proper denominator proposed by Wang is also of importance.

In the 18th century, European mathematics had already been introduced into China but there is no evidence of their influence in Wang's paper. He attributed his ideas about numeral systems with different bases to the Yi Jing 《易經》 (The Book of Changes). It is of interest to note that Leibniz also saw in the Yi-Jing the concept of binary numerals.[4] The author thinks it is quite possible that the Yi Jing played a role in Wang's work on numeral systems of variable bases.

References

1. Heng Zhai 衡齋 was Wang Lai's 汪萊 literary name.

2. Qian Bao-Cong 錢寶琮 , 'Wang Lai Heng-Zhai Suan-Xue Ping Shu'〈汪萊衡齋算學評述〉('Commentary on Wang Lai's Heng-Zhai Mathematics') Scientific Reports of the National Zhejiang 浙江 University, 2, No. 1 (1936).

3. Li Zhao-Hua 李兆華 , 'San-Liang Suan-Jing Jiao-Shi'〈參兩算經校釋〉 ('Commentary on the Mathematics of the Rules of Two and Three') Reports of Tianjin 天津 Normal University, special issue on mathematics (1985).

4. G. W. Leibniz, 'Explication de l'Arithmétique Binaire, qui se sert des seuls caractères 0 et 1, avec des Remarques sur son Utilité, et sur ce qu'elle donne les sens des anciennes Figures Chinoises de Fohy.' Mémoires de l'Académie Royale de Sciences (Paris), 3, 85 (1703).

BAO QI-SHOU 保其壽 AND HIS POLYHEDRAL HUN YUAN TU 渾圓圖*

Ko-Wei Lih 李國偉

I. Introduction

The construction of combinatorial mathematics commonly known as a 'magic square' is an arrangement of the numbers $1, 2, 3, \ldots, n^2$ into a square array such that the sum of the numbers along any row, column, or main diagonal is a fixed constant. It is usually agreed that the following diagram

4	9	2
3	5	7
8	1	6

was the first magic square ever created in the world. Its origin is shrouded in the mystical legends of ancient China. It came to be known as 'Luo Shu' 洛書 (Luo river writing). There was no clear connection between this configuration and mathematical study until the time of Yang Hui 楊輝, even though Zhen Luan 甄鸞 (sixth century) described it in his commentary to the Shu-Shu Ji-Yi 《數術記遺》 (Memoir on Some Traditions of Mathematical Art) of c. 190. In the year 1275, Yang Hui published his Xu Gu Zhai Qi Suan-Fa 《續古摘奇算法》 (Continuation of Ancient Mathematical Methods for Elucidating the Strange Properties of Numbers). The first section of the first chapter lists twenty 'zong heng tu' 縱橫圖. (Literally,

---

*A preliminary report of this research, entitled 'Some Chinese Contributions to Combinatorial Mathematics in the 19th Century,' was read at the XVIIth International Congress of History of Science, Berkeley, 1985.

'vertical and horizontal diagram.') The term actually encompasses a wider class of configurations than does the term 'magic square.' Besides magic squares, there are other, circular, arrangements of consecutive numbers 1, 2, 3, . . . n among these diagrams of 'zong heng tu.' However, in each case, a constant sum along any of the prescribed directions always emerges. Yang's book contains the earliest known reference to the term 'zong heng tu.'

Constructions of 'zong heng tu' were further pursued by mathematicians such as Din Yi-Dong 丁易東 in the Da-Yan Suo-Yin《大衍索隱》(Elucidation of the Da-Yan Method) of the Song Dynasty, Cheng Da-Wei 程大位 in the Suan-Fa Tong-Zong 《算法統宗》(Systematic Treatise on Arithmetic) of 1593, Fang Zhong-Tong 方中通 in the Shu Du Yan 《數度衍》(Generalization on Numbers and Measures) of 1661, and the more innovative Zhang Chao 張潮 in his Xin-Zhai Za-Cu《心齋雜俎》(Miscellanea of Xin Zhai) of 1670. Such activities came to an almost complete halt after the influential mathematician Mei Gu-Cheng 梅瑴成 removed all 'zong heng tu' from his 1760 edition of the popular Suan-Fa Tong-Zong. However, Cheng Da Wei's original work was transmitted to Japan and provided a great impetus to the Japanese mathematical development. The construction of magic squares and circles henceforth flourished in Japan in the seventeenth and eighteenth centuries.

When one attempts to make a circular 'zong heng tu,' the most elementary method which comes to mind is an arrangement of complementary number pairs, i.e. (1, n), (2, n-1), (3, n-2), . . . , which obviously sum to the constant n+1. From a mathematical point of view, configurations based on this method are of lesser interest. Most Chinese magic circles and notable Japanese ones by Isomura Kittoku and Seki Kowa were of this type.

II. Bao Qi-Shou and His Work

The construction of 'zong heng tu' was literally lifted to a higher dimension when Bao Qi-Shou 保其壽 finally appeared on the scene. Bao Qi-Shou lived in the latter part of the nineteenth century. All that is known about his life is documented in a 341-character biography in the mimeographed Nantong Xian Zhi《南通縣志》(Illustrated History of Nantong County) which is kept in the municipal library of Nantong City.

His parents died when he was young. He was not able to pursue the traditional career as a scholar-bureaucrat. He tried his hand at business without success. When the Boxer Rebellion started, he enlisted in the army, but he was assigned to a clerical position, which was not his inclination. Finally, he left the army and returned to his studies. His personality was not quite pleasant, and he was fond of writing satirical poems. Astronomy and mathematics were his true love. He regretted that most people shunned Occidental knowledge. He, at one time, chose gifted pupils and taught them arithmetic and algebra. He died when he was sixty-four. The year of his death is not clear.

My interest in Bao's combinatorial constructions was first aroused in 1980 when I read Li Yan's 李儼 article 'Zhong-Suan-Jia Zhi Zong Heng Tu Yan-Jiu'〈中算家之縱橫圖研究〉('A Study of Chinese Mathematicians' Magic Squares').[1] According to Li's account, Bao exhibited his diagrams in an article 'Zen Bu Suan Fa Hun Yuan Tu'〈增補算法渾圓圖〉('Augmented Solid and Spherical Mathematical Diagrams') included in his privately published book Bi-Nai-Shan Fang-Ji《碧奈山房集》(Bi-Nai Mountain Hut Records) of c. 1880. Although Li re-drew Bao's original diagrams and changed traditional numeral symbols into current numerical symbols, he gave the reader a strong impression that he had quoted Bao's entire article.

It seems that later authoritative historians of Chinese mathematics, such as Lam[2] and Needham,[3] made their comments about Bao's work based on Li's version. However, they did not fully explore the subtleties of Bao's diagrams and failed to give him proper recognition. Since the few libraries which have Bao's book in their collections are quite inaccessible, a copy of his 'hun yuan tu' article was forwarded to me for study only recently. I am not certain whether there are other parts of his book containing mathematical matters of relevance and merit. Nevertheless, Bao's own article reveals more than Li's version conveyed.

III. Zen Bu Suan-Fa Hun Yuan Tu 〈增補算法渾圓圖〉

There are two volumes to Bao's book. Figure 1 reproduces the first page of Volume Two, which is the volume containing his diagrams. The article 'Zen Bu Suan-Fa Hun Yuan Tu' occupies

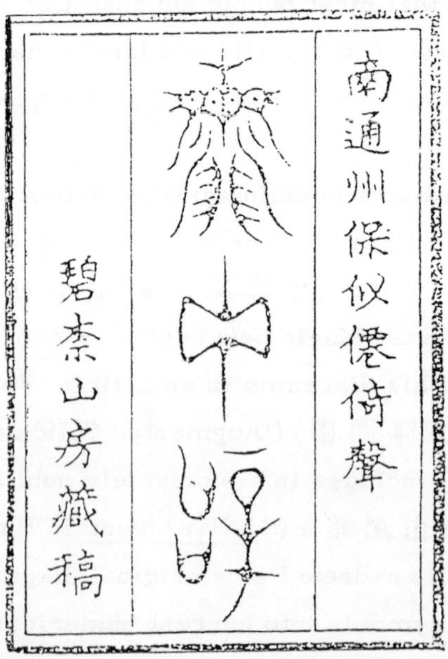

Figure 1

twenty-four pages. Since there are no page numbers in the original volume, my page count starts from the beginning of this article. I will also use 'Lx' to indicate that the same diagram appears as Figure x in Li Yan's aforementioned article. It is not feasible for me to reproduce all of the diagrams in this paper. Therefore, references to Li Yan's more accessible book should be useful to the reader.

Pages 1 and 2 contain the foreword of the article. The translation is as follows.

> There are 25 diagrams in the Xin-Zhai Za-Cu. Zhang Chao said that these were not included in the Suan-Fa Tong-Zong. They were derived along similar lines and many more could be constructed. I have not seen the Suan Fa Tong Zong. However, what Zhang Chao derived were plane diagrams. He did not realize that cubes and spheres could be of more interest. All of these diagrams have their origins in the 'Luo shu'. The ingenuity of their construction is unbelievable. Nature possesses such wonderful things and reveals them through Zhang Chao's hands and mine.

What Bao did was to put numeral labels 1, 2, 3, ... up to a suitable number on the vertices, edges, and faces of tetrahedra, cubes, octahedra, dodecahedra, icosahedra, and icosidodecahedra (a 32 sided figure with 12 pentagonal faces and 20 triangular faces) so that the sum of the vertex, edge, and face labels on a face of a given number of sides is a constant. This means that all face sums are equal except on the icosidodecahedron, which has triangular and pentagonal faces, hence two kinds of face sums. The constant sums phenomenon preserves the essence of 'zong heng tu.' However, Bao called them 'hun yuan tu' 渾圓圖 to emphasize their three dimensionality. Below are some comments on Bao's constructions.

IV. The Group of Cubes

Page 4 (L51). This is essentially an arrangement of complementary pairs (1,8), (2,7), (3,6), (4,5). Since labels are placed on vertices only, it is also called a vertex labeling.

Page 5 (L52). This is said to be an edge labeling since only edges are labeled.

Page 6 (L53). The label 2 on the top edge should be 20. Both Bao's original and Li's depiction are incorrect here. Li omitted Bao's statement that this construction is obtained by combining the previous two. From this statement we know that Bao was aware of the general method by which a new 'hun yuan tu' arises when a figure with only vertex labels is superimposed over an identical figure with only edge labels. In order to conform to the principle that the number labels cover all numbers from 1 to n without repeating, the numbers for either the edge or the vertex labels must all be increased by a constant amount equal to the last vertex or edge label, respectively.

Page 7 (L54). This labeling also appears in Needham.[3] It illustrates a second general principle that complementary pairs can be added to a vertex labeling to obtain a new 'hun yuan tu.'

V. The Group of Tetrahedra

Page 8 (L55). This is a straightforward labeling, by complementary pairs, of the edges of a tetrahedron viewed downward from the top vertex.

Page 9 (see Figure 2). Li Yan did not include this diagram in his article. Without labels, this triangular diagram is a so-called planar net representation of a tetrahedron, i.e. a tetrahedron will be obtained when the three exterior triangles on the rim are folded up and the three number 1 vertices are identified as one. Bao states that each triangular face needs its own label. This also indicates the following third general principle that Bao was aware of. A labeling is said to be of the consecutive type if the face sums form a consecutive sequence of numbers m, m+1, . . . , m+k. It is straightforward to convert a consecutive type labeling into a constant sums labeling by filling in another sequence of suitable consecutive numbers as face labels. When the numbers 5, 6, 7, 8 are removed, Figure 2 becomes a consecutive type labeling of a tetrahedron.

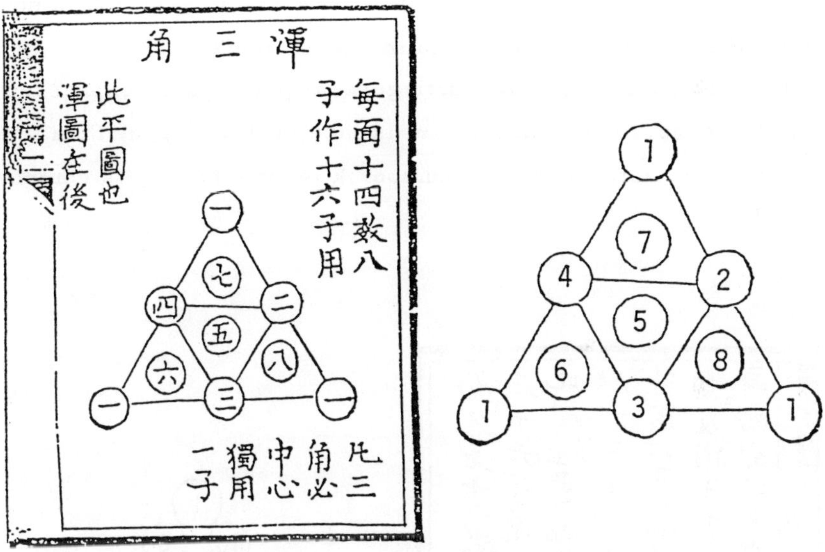

Figure 2

Page 10 (L56). This is the same labeling as Figure 2 but this figure is constructed in the form of a solid tetrahedron. The meaning of Bao's statement 'xin tong ma she pi qiu' 形同馬射皮毬 is unclear to me.

Page 11 (L57. See Figure 3). Bao states that a constant sums labeling cannot be produced if 1, 2, 3, 4 are placed on vertices and 5 to 10 are placed on edges. He also states that the face sum is not divisible by 4 if 1 to 16 are used and 1, 2, 3, 4 are placed on vertices. Since each vertex of a tetrahedron is adjacent to all other vertices, there is essentially only one unique way to consecutively label the vertices 1, 2, 3, 4. By trial and error, it is easy to see that there is no way to produce constant face sums with consecutively labeled edges. This verifies Bao's first statement. If the constant sums phenomenon is going to occur, then the total sum must be divisible by 4 when each vertex label is counted three times and each edge label is counted twice. Now $(1+2+3+4)\times 3+(5+\cdots+16)\times 2 = 282$ which is not a multiple of 4. So Bao's second claim makes sense.

Page 12 (L58, L59). A fourth general principle is expressed by Bao on this page. If a 'hun yuan tu' is labeled 1, 2, . . . , n, then a new 'hun yuan tu' results when each number k is switched to its complement n+1-k.

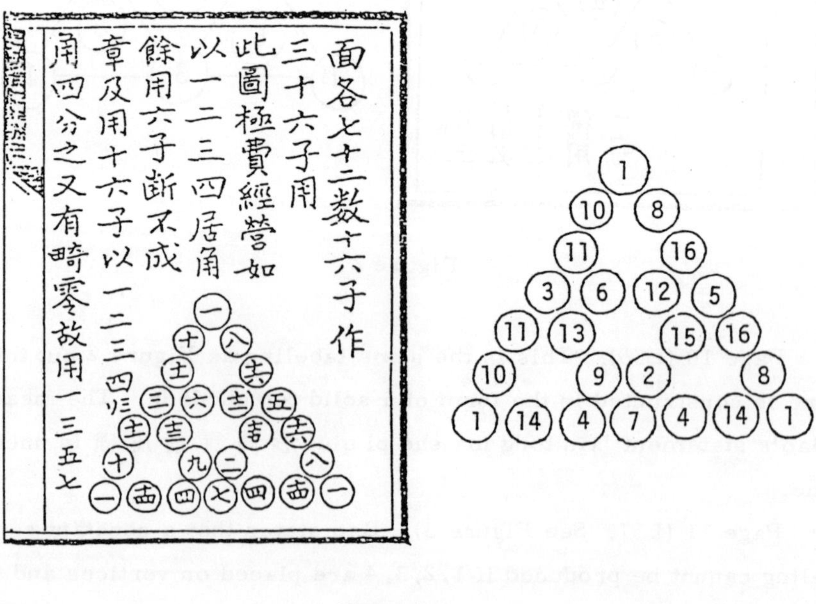

Figure 3

VI. The Group of Octahedra

Page 3 (L49, L50). Two representations of the same labeling, which is based on a consecutive vertex labeling, are exhibited here.

Page 13 (L60). The method of switching to complements is explicitly stated on this page. Another exchange of labels is also expressed here which does not appear in Li's quotation. The exchange is done by replacing 1 to 12 by 7 to 18 and 13 to 18 by 1 to 6, respectively.

Page 14 (L61). This is an edge labeling of complementary pairs. The correct face sums, when these pairs are added to the previous labeling and its complement, should be 231 and 249 instead of the stated 109 and 127.

VII. The Group of Icosahedra

Page 15 (L62. See Figure 4). The numbers 22 and 50 should

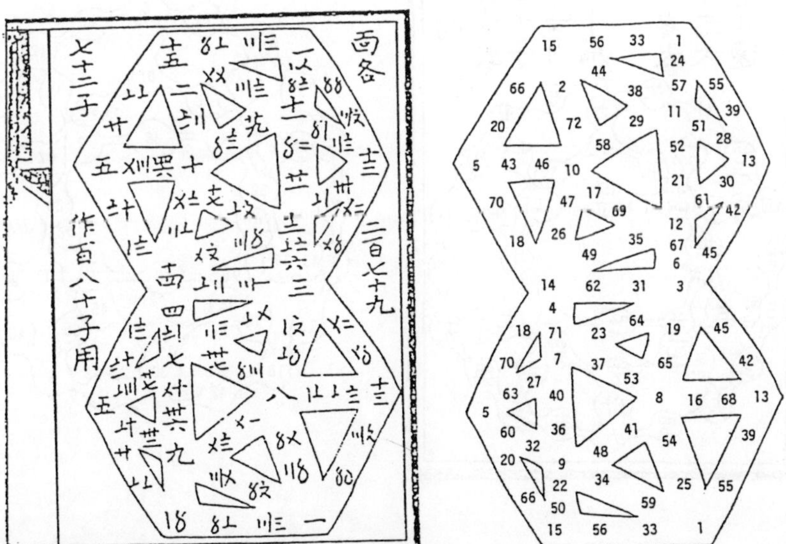

Figure 4

be placed on the edge between the labels 9 and 15. Li's version has these numbers filled in. However, his 55 between 8 and 19 is 65 on the original and the positions of 42 and 45 are interchanged.

Page 16 (L63). The number 24 between 70 and 2 is a mistake. Li's version has it changed to 80.

VIII. The Group of Dodecahedra

Page 17 (L64). The vertices are consecutively labeled.

Page 19 (L65. See Figure 5). Li's version is drawn as a planar-net representation. Bao's original diagram looks more interesting in that the labels on the back are printed upside down. An

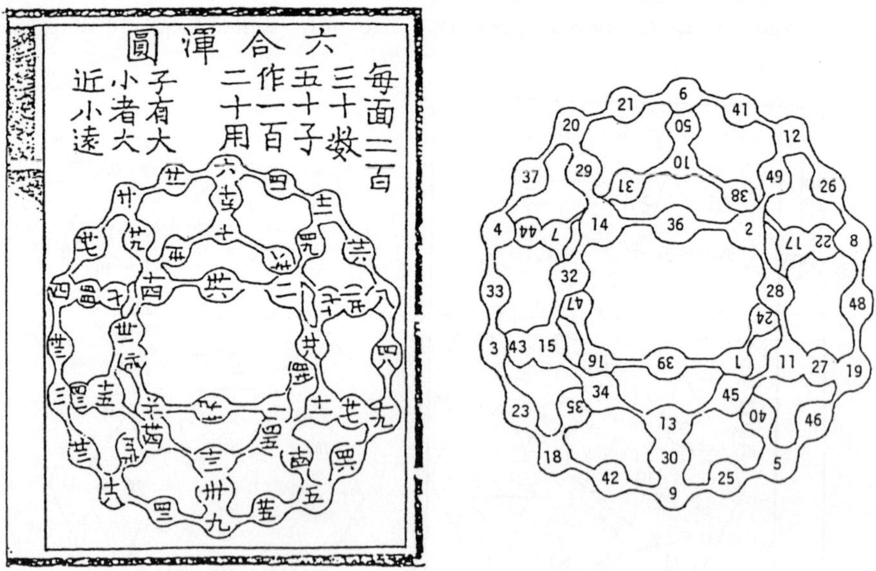

Figure 5

amazing fact about this diagram is that the vertex labels are exactly the same as those on page 17. So when the vertex labels are removed and 20 is subtracted from each edge label, a figure whose edges are consecutively labeled is formed and is complementary to the figure whose vertices are consecutively labeled on page 17 to make a constant sums labeling of both vertices and edges. This coupling method is a refined version of the third general principle used on page 9.

Page 18 (see Figure 6). This diagram is deleted in Li's version. However, it explicitly proves that a vertex labeling of the constant sums type does not exist.

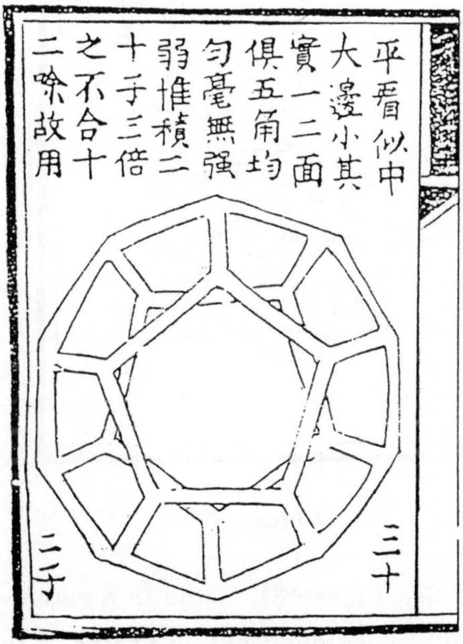

Figure 6

IX. The Group of Icosidodecahedra

Page 23 (see Figure 7). The diagram shows a ball wrapped by six intertwining ribbons. Regarding the ribbons as edges, this configuration is combinatorially the same as an icosidodecahedron. There are two types of faces, namely triangles (20) and pentagons (12). Bao uses a divisibility proof to show that a vertex labeling of the constant sums type does not exist.

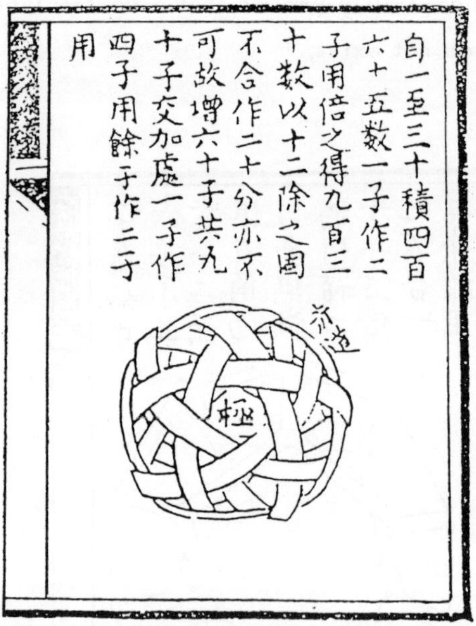

Figure 7

Page 20 (L66. See Figure 8). This is a rather complicated construction. It would be tedious work to arrive at this labeling by trial and error. Whether Bao employed some systematic method is difficult to ascertain.

Page 21 (L67). The way to get this labeling is to add 60 to each vertex and subtract 30 form each edge in the previous diagram, then

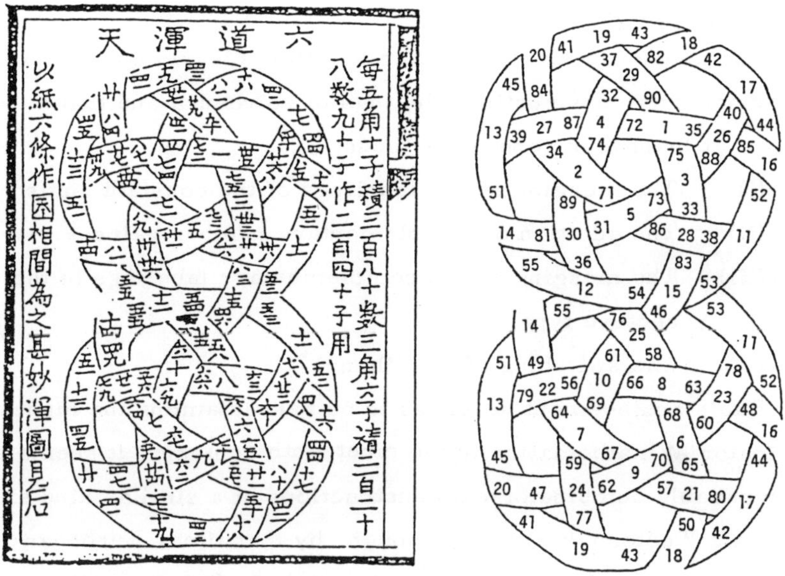

Figure 8

reflect the resulting labeling through a vertical mirror.

Page 22 (L68). This is complementary to the labeling of page 20. The number 36 should be added between 77 and 79 as in Li's version.

Page 24. This is a page of epilogue. It is translated as follows.

> Further diagrams can be constructed by the previous methods. It becomes easier to make variations when more and more numbers are used. However, it is pointless to give those illustrations when all methods have been exhaustively demonstrated.

X. Concluding Remarks

The above examination of Bao's article clearly shows that he was consciously constructing labelings of polyhedra and that he was aware of at least four general methods:

(i) The merging of two labelings of the constant sums type.

(ii) The addition of complementary number pairs.

(iii) The merging of two complementary labelings of the consectuve type.

(iv) The switching of k with n+1-k.

Bao's achievement deserves to be ranked among the highest in combinatorial mathematics in the nineteenth century. It seems that no comparable combinatorical constructions of a similar kind existed then in the Western world even though, by that time, European mathematicians had advanced far beyond their Chinese colleagues on almost every other front.

Another point of importance in Bao's constructions comes from the possibility of further generalizations. Interpreted in the modern terminology of graph theory, Bao's constructions are labelings of plane graphs, i.e. graphs drawn on the Euclidean plane such that edges do not cross each other except at vertices. Both constant sums and consecutive labelings can be defined for general plane graphs. This leads to interesting and challenging problems in graph theory. In an earlier paper,[4] I have succeeded in making such constructions for several infinite families of graphs. A consecutive vertex labeling of the icosahedron is also included in that paper. This is one of the basic labelings of platonic polyhedra missed by Bao. However, I have to leave the existence of a consecutive edge labeling complementary to it as an open question.

Acknowledgment: The author is grateful to Chen Kuei-Mei 程桂梅 for the preparation of the transcriptions of the diagrams.

References

1. Li Yan 李儼 , Zhong-Suan-Shi Lun-Cong《中算史論叢》(Collected Essays on Chinese Mathematics) (Commercial Press, Shanghai, 1935), Vol. 3, pp. 59-110.

2. Lam Lay Yong 藍麗蓉 , A Critical Study of the Yang Hui Suan Fa, A Thirteenth-century Chinese Mathematical Treatise (Singapore University Press, 1977), pp. 321-322.

3. Joseph Needham, Science and Civilisation in China (Cambridge University Press, 1959), Vol. 3, p. 60.

4. Lih Ko-Wei 李國偉 , 'On Magic Squares and Consecutive Labelings of Plane Graphs,' Utilitas Mathematica, 24, 165-197 (1983).

A STUDY OF THE MATHEMATICAL WORKS OF LI SHAN-LAN 李善蘭*

ZHAO-HUA LI 李兆華

Li Shan-Lan 李善蘭 (1811-1882) was an important mathematician of the 19th century. Among his mathematical works, the Duo-Ji Bi-Lei《垛積比類》(Mathematical Analogy of Packing and Piling Problems) of 1859 and the Fang-Yuan Chan-You《方圓闡幽》(Exposition on Squares and Circles) of 1845 are considered to be his major works.[1] In this paper, a brief discussion of the content as well as the interrelationship of these two books is presented.

I. The Duo-Ji Bi-Lei《垛積比類》

A. Description of Content

In this work, Li Shan-Lan laid emphasis on the summation of series. In Chapter I, he gave the formula

$$\sum_{r=1}^{n} \frac{1}{p!} r(r+1)(r+2) \cdots (r+p-1)$$

$$= \frac{1}{(p+1)!} n(n+1)(n+2) \cdots (n+p) \qquad (1)$$

where the r-sum actually sums over the numbers along the pth line up to the nth position in the Jia Xian 賈憲 triangle[2] (known in the West as the 'Pascal triangle') given in Figure 1. The sum gives the

*Parts of this paper were presented at the 17th International Congress of History of Science, University of California, Berkeley, July 31-August 8, 1985.

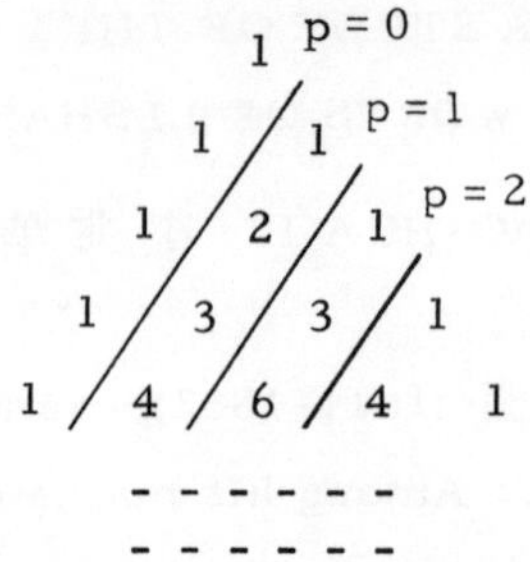

Figure 1. The Jia Xian 賈憲 triangle of the 11th century (known in the West as the 'Pascal triangle') (see Ref. 2).

number in the nth position of the (p + 1)th line. This is a well-known formula probably first given by Jia Xian 賈憲 (11th century).[3] Li took it as the basic formula for his work.

In Chapter 2, Li gave the formula for the summation of powers:

$$\sum_{r=1}^{n} r^{p+1} = \sum_{i=1}^{p+1} \left\{ A_i^p \sum_{r=1}^{n-i+1} \frac{1}{(p+1)!} r(r+1)\cdots(r+p) \right\} \tag{2}$$

where A_i^p is the number in the ith place of the pth line in the diagram shown in Figure 2a. Li pointed out that

$$A_i^p = (p-i+2)\, A_{i-1}^{p-1} + i\, A_i^{p-1} \tag{3a}$$

It follows that

$$\sum_{i=1}^{p+1} A_i^p = (p+1)! \tag{3b}$$

(a)

1					
1	1				p = 1
1	4	1			p = 2
1	11	11	1		p = 3
1	26	66	26	1	
- - - - - - - -					
- - - - - - - -					

(b)

1					
1	1				p = 1
1	4	1			p = 2
1	9	9	9		p = 3
1	16	36	16	1	
- - - - - - - -					
- - - - - - - -					

Figure 2. (a) Triangular tabulation of the coefficients of the series given by Eq. (2).
(b) Triangular tabulation of the coefficients of the series given by Eq. (4).

In Chapter 3, Li gave

$$\sum_{r=1}^{n} \left[\frac{1}{p!}\, r(r+1) \cdots (r+p-1)\right]^2$$

$$= \sum_{i=1}^{p+1} \left\{ C_i^p \sum_{r=1}^{n-i+1} \frac{1}{(2p)!}\, r(r+1) \ldots (r+2p-1) \right\} \qquad (4)$$

where C_i^p is the number in the ith place of the pth line in the diagram shown in Figure 2b. It is obvious that every number in the diagram of Figure 2b can be obtained by squaring the corresponding numbers in the Jia Xian Triangle (Figure 1).

In Chapter 4, Li gave

$$\sum_{r=1}^{n} \frac{1}{(p-1)!}\, r(r-1) \cdots (r+p-2)\, r^s$$

$$= \sum_{i=1}^{s+1} \left\{ B_i^s \sum_{r=1}^{n-i+1} \frac{1}{(p+s-1)!}\, r(r+1) \cdots (r+p+s-2) \right\} \qquad (5)$$

where B_i^s is the polynomial in the ith place of the sth line in the diagram shown in Figure 3. The diagram has the following properties:

$$B_i^s = (s-i+p)\,B_{i-1}^{s-1} + i\,B_i^{s-1} \tag{6a}$$

$$\sum_{i=1}^{s=1} B_i^s = \prod_{i=1}^{s} (p+i-1) \tag{6b}$$

Obviously, when $p=2$, Eq. (6a) reduces to Eq. (3a) and Eq. (6b) reduces to Eq. (3b).

$$1$$

$$1 \qquad (p-1) \qquad\qquad s=1$$

$$1 \qquad 1+3(p-1) \qquad (p-1)^2 \qquad\qquad s=2$$

$$1 \qquad 4+7(p-1) \qquad 1+4(p-1)+6(p-1)^2 \qquad (p-1)^3 \qquad\qquad s=3$$

- - - - - - - - -
- - - - - - - - -

Figure 3. Triangular tabulation of the polynomial coefficients of Eq. (5).

In addition, Li further generalized formulas given by Eqs. (2), (4) and (5). We have from Eq. (2)

$$\sum_{r=1}^{n}{}^{k}\, r^{p+1} = \sum_{i=1}^{p+1} \left\{ A_i^p \sum_{r=1}^{n-i+1} \frac{1}{(p+k)!}\, r(r+1)\ldots(r+p+k-1) \right\} \tag{7}$$

where A_i^p is the same as in Eq. (2) satisfying Eq. (3) and the symbol Σ^k stands for summation k times. For example, when $p=1$, from Eq. (2), we have

$$\sum_{r=1}^{n} r^2 = \sum_{i=1}^{2} \left\{ A_i^1 \sum_{r=1}^{n-i+1} \frac{1}{2!} r(r+1) \right\}$$

$$= A_1^1 \sum_{r=1}^{n} \frac{1}{2!} r(r+1) + A_2^1 \sum_{r=1}^{n-1} \frac{1}{2!} r(r+1)$$

$$= \sum_{r=1}^{n} \frac{1}{2!} r(r+1) + \sum_{r=1}^{n-1} \frac{1}{2!} r(r+1)$$

$$= \frac{1}{3!} n(n+1)(n+2) + \frac{1}{3!} (n-1) n(n+1)$$

$$= \frac{1}{3!} n(n+1)(2n+1) .$$

Following Li's procedure, we define

$$\sum_{r=1}^{n}{}^{2} r^2 = \sum_{r=1}^{n} \frac{1}{3!} r(r+1)(2r+1).$$

By using the above result, we have

$$\sum_{r=1}^{n} \frac{1}{3!} r(r+1)(2r+1)$$

$$= \sum_{r=1}^{n} \left[\frac{1}{3!} r(r+1)(r+2) + \frac{1}{3!} (r-1) r(r+1)\right]$$

$$= \sum_{r=1}^{n} \frac{1}{3!} r(r+1)(r+2) + \sum_{r=1}^{n} \frac{1}{3!} (r-1) r(r+1)$$

$$= \sum_{r=1}^{n} \frac{1}{3!} r(r+1)(r+2) + \sum_{r=1}^{n-1} \frac{1}{3!} r(r+1)(r+2)$$

$$= \sum_{i=1}^{2} \left[A_i^1 \sum_{r=1}^{n-i+1} \frac{1}{3!} r(r+1)(r+2) \right]$$

that is

$$\sum_{r=1}^{n} {}^2 \, r^2 = \sum_{i=1}^{2} \left[A_i^1 \sum_{r=1}^{n-i+1} \frac{1}{3!} r(r+1)(r+2) \right] .$$

Similarly, we have

$$\sum_{r=1}^{n} {}^3 \, r^2 = \sum_{i=1}^{2} \left[A_i^1 \sum_{r=1}^{n-i+1} \frac{1}{4!} r(r+1)(r+2)(r+3) \right]$$

and so forth.

Thus, from Eq. (4), we have the general formula

$$\sum_{r=1}^{n} {}^k \left[\frac{1}{p!} r(r+1) \cdots (r+p-1) \right]^2$$

$$= \sum_{i=1}^{p+1} \left\{ C_i^p \sum_{r=1}^{n-i+1} \frac{1}{(2p+k-1)!} r(r+1) \cdots (r+2p+k-2) \right\} \quad (8)$$

where C_i^p is the same as in Eq. (4). Finally from Eq. (5), we have

$$\sum_{r=1}^{n} {}^k \frac{1}{(p-1)!} r(r+1) \cdots (r+p-2) r^s$$

$$= \sum_{i=1}^{s+1} \left\{ B_i^s \sum_{r=1}^{n-i+1} \frac{1}{(p+k+s-2)!} r(r+1) \cdots (r+p+k+s-3) \right\} \quad (9)$$

where B_i^s is the same as in Eq. (5), satisfying Eq. (6).

B. Derivation of Li's Summation Formulas

In the preface to his work, Li mentioned Guo Shou-Jing 郭守敬 (1231-1361), Zhu Shi-Jie 朱世傑 of the late 13th and early 14th centuries, Wang Lai 汪萊 (1768-1313) and Dong You-Cheng 董祐誠 (1791-1823) and their work on packing and piling problems in mathematics. It is known that Zhu Shi-Jie provided an explicit derivation of $\Sigma^2 r^3$ and that formula (1) and the derivation of $\Sigma^k r^2$ were given by Wang Lai and Dong Yon-Cheng, respectively. This suggests that Li's summation formulas were derived from Eq. (1) within the framework of the 'tian yuan shu' 天元術 ('the method of celestial elements'), the traditional Chinese algebra.

As an example, I will show how the formula given by Eq. (2) is derived from Eq. (1). We have from Eq. (1)

$$\sum_{r=1}^{n} \frac{1}{p!} r(r+1)\ldots(r+p-1) = \sum_{r=1}^{n} \frac{1}{(p+1)!} r(r+1)\cdots(r+p) - \sum_{r=1}^{n-1} \frac{1}{(p+1)!} r(r+1)\cdots(r+p) \tag{10}$$

Taking $p = 1$, Eq. (10) reduces to the form

$$\sum_{r=1}^{n} r = \sum_{r=1}^{n} \frac{1}{2!} r(r+1) - \sum_{r=1}^{n-1} \frac{1}{2!} r(r+1) \tag{11}$$

By rearranging the identity

$$\sum_{r=1}^{n} \frac{1}{2!} r(r+1) = \frac{1}{2!}\left(\sum_{r=1}^{n} r^2 + \sum_{r=1}^{n} r\right)$$

in the form

$$\sum_{r=1}^{n} r^2 = 2!\sum_{r=1}^{n} \frac{1}{2!} r(r+1) - \sum_{r=1}^{n} r$$

we obtain, with the help of Eq. (1),

$$\sum_{r=1}^{n} r^2 = \sum_{r=1}^{n} \frac{1}{2!} r(r+1) + \sum_{r=1}^{n-1} \frac{1}{2!} r(r+1) \qquad (12)$$

Similarly, we obtain

$$\sum_{r=1}^{n} r^3 = \sum_{r=1}^{n} \frac{1}{3!} r(r+1)(r+2)$$

$$+ 4\sum_{r=1}^{n-1} \frac{1}{3!} r(r+1)(r+2) + \sum_{r=1}^{n-2} \frac{1}{3!} r(r+1)(r+2)$$

$$\sum_{r=1}^{n} r^4 = \sum_{r=1}^{n} \frac{1}{4!} r(r+1)(r+2)(r+3)$$

$$+ 11\sum_{r=1}^{n-1} \frac{1}{4!} r(r+1)(r+2)(r+3)$$

$$+ 11\sum_{r=1}^{n-2} \frac{1}{4!} r(r+1)(r+2)(r+3)$$

$$+ \sum_{n=1}^{n-2} \frac{1}{4!} r(r+1)(r+2)(r+3) \qquad (14)$$

and so on. The coefficients 1, 1; 1, 4, 1; 1, 11, 11, 1; etc. are known as Stirling's numbers in modern combinatorial theory.

Having obtained Eq. (2), we can derive formulas given by

Eqs. (4) and (5) by this procedure. Similarly, formulas given by Eqs. (7), (8) and (9) can be obtained by using Eqs. (2), (4) and (5).

II. The Fang-Yuan Shan-You 《方圓闡幽》

A. Description of Content

In this book, Li gave the definition of a pyramid with curved surfaces and its properties. He derived the formula for its volume and provided examples of its application.

According to Li, a curved-surface pyramid is made up of 'layers' of squares. If the rth layer's area is r^p with r and p being natural numbers, the pyramid is called a 'curved-surface pyramid of order p' (see Figure 4). Let the distance from the top point to the rth layer be x and the ratio between r and x be α. Then the x dependence of the rth layer's area A can be made explicit.

$$A(x) = (\alpha x)^p, \quad \text{with} \quad \alpha x = r$$

when x = h, we have

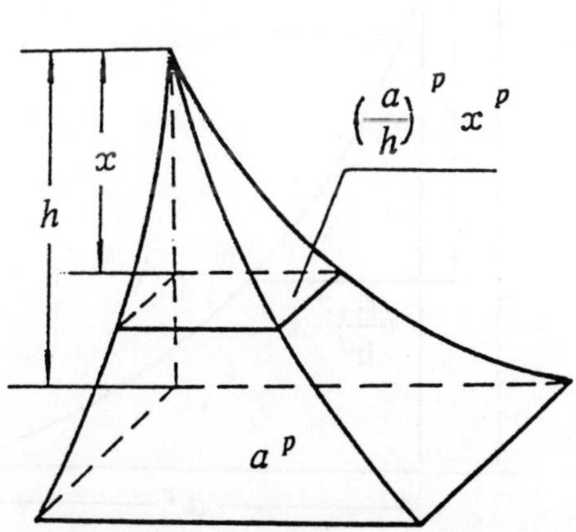

Figure 4. A curved-surface pyramid of order p with height h and base area a^p.

$$A(h) = (\alpha h)^p = a^p .$$

It then follows that

$$\alpha^p = \left(\frac{a}{h}\right)^p$$

and we have for the area of the rth layer the formula

$$A(x) = \left(\frac{a}{h}\right)^p x^p \tag{15}$$

The pyramid's volume is obtained by the accumulation of the areas. Li gave the formula

$$V = \frac{1}{p+1} a^p h \tag{16}$$

Li pointed out that each layer's area of the pyramid can be represented by a line segment so that the pyramid's volume can be represented by a curvilinear triangle as shown in Figure 5. The line segment

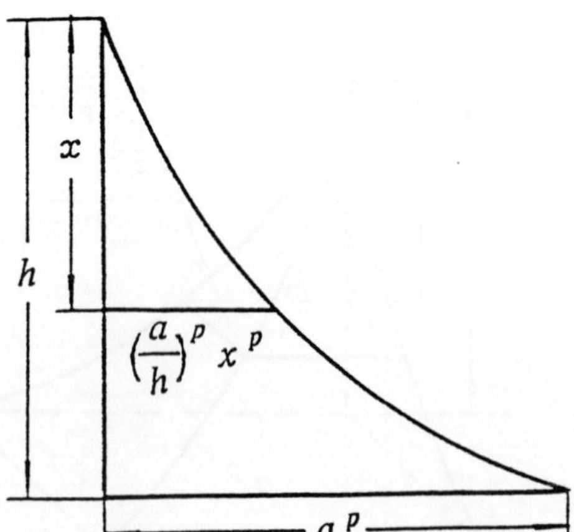

Figure 5. A curvilinear-triangle representation of a curved-surface pyramid of order p (see Figure 4).

representing the width of the curvilinear triangle at a distance x from the top point of the triangle is taken to be

$$y = \left(\frac{a}{h}\right)^p x^p \tag{17}$$

which is exactly equal to the area of the layer located at x distance from the top point of the pyramid (see Figure 4). As x increases to the value h, the line segment takes the value a^p which is the base area of the pyramid. The area of the curvilinear triangle S is given by

$$S = \frac{1}{p+1} a^p h \tag{18}$$

exactly equal to the volume of the corresponding pyramid.

Li further pointed out that the area S of a curvilinear triangle can be expressed as the sum of curvilinear area strips in the form

$$S = \sum_i S_i$$

where S_i is the ith area strip to be evaluated by Eq. (16). To demonstrate this, Li considered the example of expanding π. Shown in Figure 6 is a quarter of a circle of unit radius and its circumscribed square. In the figure, OACB is a square with side 1 unit, and OAQB is one fourth of a circle with radius 1 unit. Li showed that an expansion for π can be obtained by expressing the curvilinear triangle ACB as the sum of curvilinear area trips

$$S_{ACB} = S_{ACC_1} + S_{AC_1C_2} + S_{AC_2C_3} + \cdots \tag{19}$$

Li gave for the bases of these strips the following values:

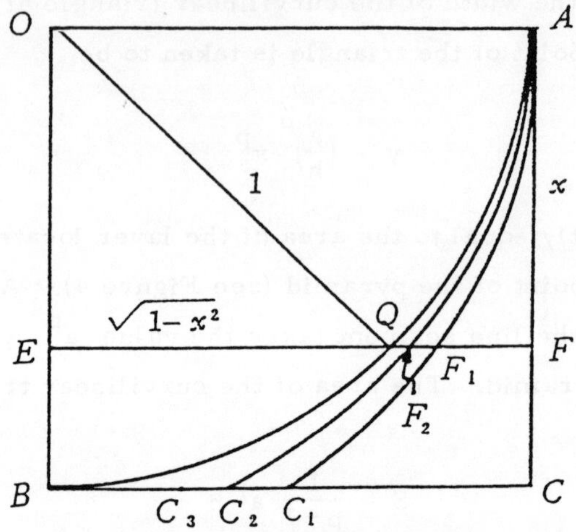

Figure 6. One quarter of a unit circle with its circumscribed square. The curvilinear triangle ACB is 1/4 of the difference in area between the circumscribed square and the circle.

$$\begin{aligned} CC_1 &= \frac{1}{2} \\ C_1C_2 &= \frac{1}{2} \times \frac{1}{4} \\ C_2C_3 &= \frac{1}{2} \times \frac{3}{4} \times \frac{1}{6} \\ C_3C_4 &= \frac{1}{2} \times \frac{3}{4} \times \frac{5}{6} \times \frac{1}{8} \end{aligned} \tag{20}$$

etc.

These strips progress with order $p = 2n$, $n = 1, 2, 3, \ldots$.

Making use of Eq. (18), he obtained, with $h = 1$ unit,

$$S = \frac{1}{2 \times 3} + \frac{1}{2 \times 4} \times \frac{1}{5} + \frac{1 \times 3}{2 \times 4 \times 6} \times \frac{1}{7} + \frac{1 \times 3 \times 5}{2 \times 4 \times 6} \times \frac{1}{9} + \cdots$$

With the radius taken to be 1 unit, the area of the circle is π (units)^2 which is 4 times the difference between the area of the unit square OACB and the curvilinear area S. Li then obtained his expansion for π:

$$\pi = 4 - 4S$$

$$= 4 - 4\left(\frac{1}{2\times 3} + \frac{1}{2\times 4}\times\frac{1}{5} + \frac{1\times 3}{2\times 4\times 6}\times\frac{1}{7} + \cdots\right) \qquad (21)$$

B. Exposition of Li's Method

Li provided no explanation of how the bases of the curvilinear strips are obtained. Examining Figure 6, it is seen that line EF is parallel to BC and intersects with the curvilinear lines at points F_1, F_2, F_3, ..., Q. Let AF = x, then $EQ = \sqrt{1 - x^2}$, and

$$FQ = 1 - \sqrt{1 - x^2}$$

And from Eq. (17) one obtains

$$FF_1 = y_1 = \frac{1}{2}x^2$$

$$FF_2 = y_2 = \frac{1}{2}x^2 + \frac{1}{2\times 4}x^4$$

$$FF_3 = y_3 = \frac{1}{2}x^2 + \frac{1}{2\times 4}x^4 + \frac{1\times 3}{2\times 4\times 6}x^6$$

etc.

then it is clear that Li's method of expanding S is equivalent to

$$S = \int_0^1 (1 - \sqrt{1 - x^2})\,dx$$

$$= \int_0^1 y_1\,dx + \left(\int_0^1 y_2\,dx - \int_0^1 y_1\,dx\right) + \left(\int_0^1 y_3\,dx - \int_0^1 y_2\,dx\right)$$

$$+ \left(\int_0^1 y_4\,dx - \int_0^1 y_3\,dx\right) + \cdots$$

$$= \int_0^1 \frac{1}{2} x^2\,dx + \int_0^1 \frac{1}{2\times 4} x^4\,dx + \int_0^1 \frac{1\times 3}{2\times 4\times 6} x^6\,dx$$

$$+ \int_0^1 \frac{1\times 3\times 5}{2\times 4\times 6\times 8} x^8\,dx + \cdots$$

$$= \frac{1}{2}\times\frac{1}{3} + \frac{1}{2\times 4}\times\frac{1}{5} + \frac{1\times 3}{2\times 4\times 6}\times\frac{1}{7} + \frac{1\times 3\times 5}{2\times 4\times 6\times 8}\times\frac{1}{9} + \cdots$$

Obviously, between Eqs. (15) and (16) we have the integration relation

$$V = \int_0^h \left(\frac{a}{h}\right)^p x^p\,dx = \frac{1}{p+1}\,a^p h$$

Li also assumed that the order of summation and integration is interchangeable.

C. The Relationship between the Curved Surface Pyramid and Piling Problems

In the Duo-Ji Bi-Lei, illustrations are given for the series. The illustrations for the series given by Eq. (2), for example, are reproduced in Figure 7. According to Li's explanation, the layers in Figures 7(a), 7(b) and 7(c) represent 'line segments,' 'planes,' and 'volumes,' respectively, and they are mutually transformable.

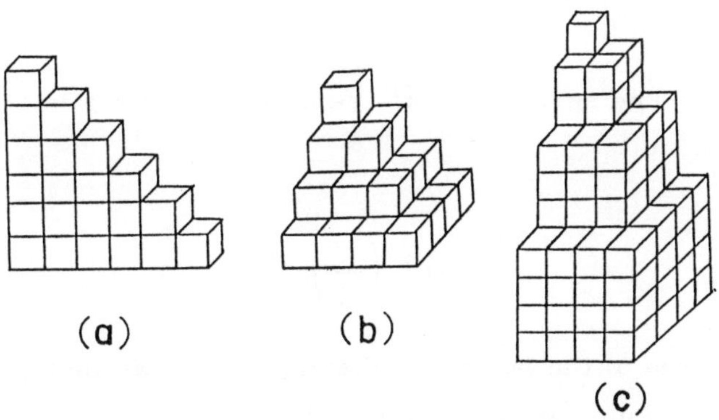

Figure 7. Illustrations for the series $\sum_{r=1}^{n} r^{p+1}$ [see Eq. (2)].

(a) $p = 0$, $\sum_{r=1}^{n} r$; (b) $p = 1$, $\sum_{r=1}^{n} r^2$; (c) $p = 2$, $\sum_{r=1}^{n} r^3$.

If one considers r^p to be one planar layer, then the summation of r^p , i.e. the series, becomes a pile of planar layers as shown in Figure 8. Let the height of the pile of planar layers be h, the area of the base be a_p^p and the distance from the top of the first layer be x. Then we have for the base area of the rth layer

$$f(x) = r^p = (\alpha x)^p$$

where $\alpha = r/x$. When $x = h$, $f(h) = a_p^p$ and then we have

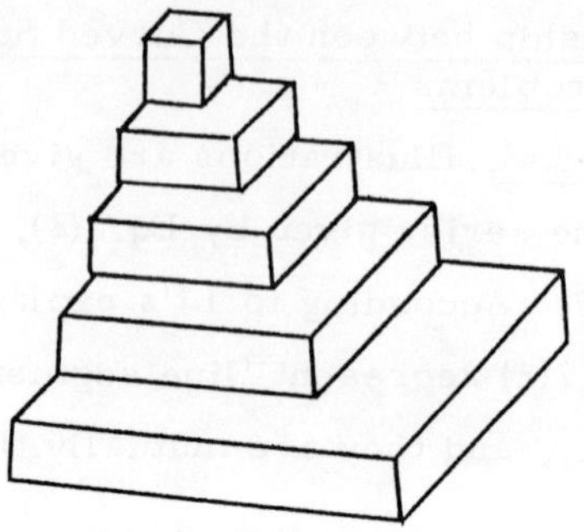

Figure 8. The series $\sum_{r=1}^{n} r^p$ represented as a pile of planes.

$$f(x) = \left(\frac{a_p}{h}\right)^p x^p$$

This is the area of the layer at a distance x from the top of the first layer. [Compare with Eq. (15).]

If there are n layers in Figure 8, the thickness of each layer is then h/n and x = r h/n). The base area of the rth layer is

$$f\left(r\frac{h}{n}\right) = \left(\frac{a_p}{h}\right)^p \left(r\frac{h}{n}\right)^p = a_p^p \left(\frac{r}{n}\right)^p$$

and the volume of the rth layer is

$$\frac{h}{n} a_p^p \left(\frac{r}{n}\right)^p = \frac{a_p^p h}{n^{p+1}} r^p$$

It is then obvious that the volume of the pile of layers given in Figure 8 is

$$V_p = \sum_{r=1}^{n} \frac{a_p^p h}{n^{p+1}} r^p = \frac{a_p^p h}{n^{p+1}} \sum_{r=1}^{n} r^p$$

From the formula given by Eq. (2), we obtain

$$V_p = \frac{a_p^p h}{n^{p+1}} \sum_{i=1}^{p} \left\{ A_i^{p-1} \sum_{r=1}^{n-i+1} \frac{1}{p!} r(r+1) \cdots (r+p-1) \right\}$$

Li reasoned that by letting the thickness of each layer be as thin as a paper, i.e. letting $h/n \to 0$, or $n \to \infty$, then

$$V = \lim_{n \to \infty} V_p = \frac{a_p^p h}{(p+1)!} \sum_{i=1}^{p} A_i^{p-1}$$

$$= \frac{a^p h}{(p+1)} p! = \frac{a^p h}{p+1}$$

Thus, the above figure changes into a curved surface pyramid.

It is seen that the method for the curved surface pyramid invented by Li Sham-Lan made use of the concept of limit and traditional Chinese piling problems. The method contains the essential features of integration but is different from the European approach. Li's work demonstrates the viability of the Chinese approach to mathematics.

References

1. Thirteen of Li Shan-Lan's 李善蘭 works were selected and engraved in 1867 as a collection entitled Ze-Gu Xi Zhai Suan-Xue《則古昔齋算學》(Mathematics Based on Traditional School).

2. The earliest extant reference of Jia Xian's work on the triangular tabulation of binominal coefficients is found in the Xiang-Jie Jiu-Zhang Suan-Fa《詳解九章算法》(A Detailed Analysis of the Mathematics in the 'Nine Chapters') of 1261 by Yang Hui 楊輝. See Yong-Le Da-Dian《永樂大典》(Great Encyclopaedia of the Yong-Le Reign), vol. 16344.

3. Qian Bao-Cong 錢寶琮, Zhong-Guo Shu-Xue Shi《中國數學史》(History of Chinese Mathematics) (Science Publ., Beijing, 1964).

SCIENCE

THE APPLICATION OF HISTORIC ASTRONOMICAL RECORDS TO ASTROPHYSICAL PROBLEMS

Ze-Zong Xi 席澤宗

I. Stellar Evolution

V. F. Weisskopf, a former president of the American Council of Atomic Energy, has said that in the history of mankind there are two 4th of Julys which should be recorded in the annals. One is in 1776 when the United States of America was founded, another is in 1054 when Chinese and Japanese astronomers recorded the supernova explosion in Taurus.[1] In 1921 the Swedish astronomer K. Lundmark first noted that the position of a supernova recorded in 1054 was very close to that of the Crab Nebula, and suspected that the two were related.[2] Soon after that, calculations for the expansion speed of the Crab Nebula by E. Hubble showed that the explosion had taken place 900 years before, consistent with the year of the recording of the supernova explosion in 1054.[3] In 1942, J. H. Oort and J. J. Duyvendark confirmed that the Crab Nebula was a remnant of the explosion in 1054 and that the explosion was that of a supernova rather than of a nova.[4-6]

The identification of the 1054 supernova with the Crab Nebula is an important piece of evidence for the value of historical records to astrophysics and the Crab Nebula is considered the Rosetta stone of astrophysics. Most of what we know about the origin of cosmic rays, synchrotron radiation and heavy elements derives from our knowledge of the Crab Nebula, and much of what we can deduce from observations of the Crab Nebula is aided by the records of the supernova explosion in 1054. In addition to this, historical records of galactic supernovae can give information on the frequency of such outbursts and on the development of their remnants. For example, from Sedov's equation[7]

$$D = 4.3 \times 10^{-11} (E_o/n)^{1/5} t^{2/5} \qquad (1)$$

where D (pc) is the diameter of the shock wave preceding the expanding shell of swept-up interstellar material, t (yr) is the time elapsed since the explosion, E_o (erg) is the energy released in the outburst, and n (cm^{-3}) is the number density of hydrogen atoms in the interstellar medium. If D is known from observation and t from historical records, we can obtain E_o/n. In 1976 Clark and Caswell concluded that the best estimate was $E_o/n \approx 5 \times 10^{51}$ erg cm^3.[8]

Most early recorded observations of supernovae were made in China, Japan and Korea. In ancient and medieval Europe and the Arab lands there seems to have been little interest in such phenomena. In The Historical Supernovae written by D. H. Clark and F. R. Stephenson in 1977, seven supernovae are listed (see Table I).

Table I. Supernovae in Recorded History

	Year	Constellation	Magnitude	Duration	Records
1	+185	Centaurus	-8	20 months	C
2	393	Scorpius	-1	8 months	C
3	1006	Lupus	-9.5	several years	A, C, J, E
4	1054	Taurus	-5	22 months	C, J
5	1181	Cassiopeia	0	6 months	C, J
6	1572	Cassiopeia	-4	18 months	C, K, E
7	1604	Ophiuchus	-2.5	12 months	C, K, E

C: Chinese, A: Arabic, J: Japanese,
E: European, K: Korean.

Among them the supernova of 1006 is the only one known to have been recorded in both European and Arabian literature before the Renaissance. It was also carefully observed by astronomers in China and Japan. It was the most brilliant supernova ever recorded and was observed for several years. There are various opinions about which object represents its remannts: NGC 5882,[9] PKS 1459-42,[10] G 327.6 + 14.5[11] or PKS 1527-42.[12] Startlingly, re-explosion of the supernova probably took place on May 16, 1016 according to Chinese records.[13] If this were true, it would support the theory of abnormal neutron stars, put forward by Fang Li-Zhi 方勵之 and Qu Qin-Yue 曲欽岳. According to the theory, as the remnant of a supernova, a neutron star may, under certain conditions, transfer its phase from normal to abnormal by releasing vast quantities of energy, on the order of 10^{52} ergs (equivalent to a supernova outburst).[14]

In 1955 the author compiled a new catalogue of ancient novae[15] and in 1965 with his colleague Bo Shu-Ren 薄樹人 revised and republished it as an appendix to 'Zhong Chao Ri San-Guo Gu-Dai de Xin-Xing he Chao-Xin-Xing Ji-Lu Ji Qi Zai She-Dian Tian-Wen-Xue Zhong de Yi-Yi' 〈中朝日三國古代的新星和超新星記錄及其在射電天文學中的意義〉 ('Nova and Supernova Records of Chinese, Japanese and Korean Annals and Their Significance in Radio Astronomy').[16] These catalogues have aroused much interest outside of China and have been translated into English and Russian. The availability of these catalogues has made possible an increase in the use of Chinese records for astrophysical research as evidenced by the fact that in the past twenty years more than 1000 articles have cited them. In 1980 G. G. C. Palumbo, G. K. Miley and P. Schiavo Campo selected seven objects from the latter catalogue, and observed the field $1.5° \times 1.5°$ around each object with the Westerbork radio telescope in Holland in an attempt to discover non-thermal radio sources. In spite of no diffusion radio emission being detected, they suggested that the investigation be continued.[17]

Based on one entry of the event occurred on September 10, 1408 that he found in Xi's catalogue (1955)[15] and eight additional references to this event found by Zhang Wei-Feng 莊威鳳 et al. in the Di-Fang Zhi《地方誌》(Local Chronicles) and the Ming Shi Lu《明實錄》(Veritable Records of the Ming Dynasty), Li Qi-Bin 李啓斌 concluded in 1978 that the event was a supernova explosion which gave rise to the progenitor of Cyg X-1, a prime black hole candidate.[18] In addition to the Chinese sightings, Imaeda and T. Kiang 江濤 pointed out that the 'guest star' was also seen in Japan. The first Japanese account is dated July 14, 1408, which precedes the first Chinese record by 58 days. Combining the Chinese and Japanese dates, we can conclude that the event of 1408 was visible for at least 102 days. The long duration indicates that the event was a supernova.[19] However, the position deduced for the supernova of 1408 places it near two astronomical curiosities, Cyg X-1 and CTB 80. There is some confusion over the interpretation of the original Chinese as to whether the location should be to the southeast of or in the southeast of Niandao 輦道 (17 Cyg). The latter interpretation would favor Cyg X-1 and the former, CTB 80.[20] In 1984 Wang Zhen-Ru 汪珍如 and F. D. Seward re-analyzed both the low and high resolution images of CTB 80 from the Einstein satellite. The result supports the view that CTB 80 is the remnant of the supernova explosion in 1408.[21] Including CTB 80, Wang et al. have identified eight supernova remnants with ancient guest stars[22,23] (see Table II).

Cassiopeia A, the strongest radio source in the sky, was considered to be a supernova remnant as early as the 1950s, before any record of it had been found.[24] In 1968 the Korean scholar Chu Sun-Il suggested that it was a remnant of the supernova recorded as being

Table II. Eight Pairs of Identification between AGS and SNR.

	Year	SNR	Records	No. in Xi and Bo's Catalogue
1	-532	G74.9 + 1.2 (CTB 87)	C	2
2	-134	G322.4 - 0.4 (RCW 103)	C, G	4
3	-48	G21.5 - 0.9	C	7
4	125	G31.9 + 0.0 (3C391)	C	13
5	421	G299.0 + 1.8 (MSH 11-54)	C	30
6	437	2CG195 + 4 (Geminga)	C	32
7	1408	G68.2 + 2.6 (CTB 80)	C, J	78 (Xi's catalogue)
8	1523	G29.7 - 0.3 (Kes 75)	C	78

C: Chinese, G: Greece, J: Japan.

observable to the west of the first star of Wang Liang 王良 (located on the west side of β Cas) from December 6, 1592 to March 5, 1593 in the Korean books the Lee-Jo Si-Lok《李朝實錄》(Chronicles of the Lee Dynasty) and the Chung-Bo Mun-Hun Bi-Go《增補文獻備考》(Supplemental Examination of Literature) for their positions seem to be very close.[25] But S. F. Gull pointed out that the preferred age of the remnant is much less than this.[26] Based on study of the proper motion of the filamentary structure in this optical remnant, Sidney Van den Bergh and W. W. Dodd placed its occurrence in the year 1667 ± 8.[27] In 1980 a new discovery made by W. B. Ashworth revealed that the first British Astronomer Royal, John Flamsteed, had recorded the event after all, fixing the actual year of the supernova explosion as 1680.[28] K. Brecher immediately wrote a paper on the finding by combining the actual date of the event with the extrapolated date to determine the mass ejected from the supernova, which turned out to be ten times greater than that of the Sun.[29] However, R. P. Broughton and K. W. Kamper denied Ashworth's discovery quickly and respectively.[30,31] So up to now the identification of Cas A with a supernova remnant has not yet been solved.

Chinese annals record not only the sudden changes in the process of stellar evolution such as supernova explosions, but also the gradual changes. In the astronomical chapter of the Shi Ji 《史記》 (Historical Records) by Sima Qian 司馬遷 about the -1st century, it is said: 'Yellow is like Betelgeuse (α Ori),' but Ptolemy included it in the list of red stars in his Almagest which was written about +150. Normal stellar evolution, even of a supergiant, from yellow to red in 250 years seems unlikely. However, a shell of gas and dust of about a 30" radius (about 5000 AU at the 190 pc distance to the star) appears to be moving away from α Ori with a present speed of about 10 km/s. Brecher has explored the possibility that the ejection of the shell occurred about 2700 years ago, temporarily giving rise to a smaller, hotter and whiter photosphere which then readjusted to nearly its original equilibrium state between the times of the writing of the Shi Ji and the Almagest.[32,33]

II. Solar Activity

Another use for ancient astronomical records in astrophysics in recent years is in discussing the periodicity of solar activity. In 1976 J. A. Eddy again strengthened the plausibility of the Maunder Minimum,[34] in which the sunspot cycle essentially vanished between 1645 and 1715, by a re-analysis of contemporary literature detailing sunspot counts, auroral records, and observations of the Sun at eclipse, plus indirect evidence from C^{14} in tree rings. Eddy even claimed that there was insufficient evidence to establish whether the 11-year cycle existed before the onset of the Maunder Minimum and after the introduction of the telescope.[35] Eddy's claim caused a sensation. If it were true, solar physics would need rewriting. The only way to solve this problem is to study the historical sunspot records of China, Japan and Korea, because the Orient was free from the stranglehold of the Aristotelian doctrine of celestial perfection, and sunspots were recorded from very early times.

To accept Eddy's challenge, the Yunnan 雲南 Astronomical Observatory sorted out more than 100 records of sunspots from Chinese annals and through self-correlation analysis showed that the 11 year cycle has existed for the past 2000 years.[36, 37] At the same time, using Chinese historical sunspot records as well as auroral records from -165 to 1884, Zou Yi-Xin 鄒儀新 found that the weighted mean cycle of solar activity over the past 2000 years was 10.42 ± 0.19 years and that it was 10.54 ± 0.62 years within the Maunder Minimum. The two cycles coincided.[38]

In 1979 Xu Zhen-Tao 徐振韜 and his wife Jiang Yao-Tiao 蔣窈窕 published a paper entitled 'Cong Zhong Guo De-Fang-Zhi Zhong Tai-Yang Hei-Zi Ji-Lu Kan Shi-Qi Shi-Ji de Tai-Yang Huo-Dong' 〈從中國地方志中太陽黑子記錄看17世紀的太陽活動〉('The Solar Activity of the 17th Century Based on Sunspot Records in the Local Chronicles of China') in which they listed 21 new naked-eye sunspot records in the 17th century and six of them were within the Maunder Minimum. The study of Xu and Jiang showed that the 11 year cycle still existed within the Maunder Minimum.[39]

It is necessary to point out that in the paper ('Sunspot Records in China, Korea and Japan') by S. Kanda (1932) there is not one sunspot from 1639 to 1720.[40] Kanda's paper is one of Eddy's bases. Now Xu and Jiang have restored six occurrences of sunspots to the Maunder Minimum, and consequently cause the original basis of Eddy's idea to lose some of its validity. Therefore, their paper attracted much interest. But J. A. Eddy considered that when weighted against more than 600 sunspots reported telescopically between 1645 and 1715, these six new spots, if real, add little that is new. In this sense these new data support rather than refute his opinion, and provide stronger evidence for secular solar variability.[41]

In 1980 Dai Nian-Zu 戴念祖 and Chen Mei-Dong 陳美東 published 'Li-Shi Shang de Bei-Ji-Guang yu Tai-Yang Huo-Dong' 〈歷史上的北極光與太陽活動〉 ('The Aurora Borealis and Solar Activity in History') as well as 'Zhong Chao Ri Li-Shi Shang Cong Chuan-Shuo Shi-Dai Dao Gong-Yuan 1747 Nian de Bei-Ji-Guang Nian-Biao' 〈中朝日歷史上從傳說時代到公元1747年的北極光年表〉('A Chronological Table of the Aurora Borealis in Chinese, Korean and Japanese History from the Legendary Period to 1747').[42] Using 929 items of aurorae in an all-around way they negated the second basis of Eddy's idea. They considered that from -217 to 1749 there were 180 peak years of solar activity, and that the mean cycle was equal to 11 years. The auroral records also indicated that there were several minima of solar activity in history, and that the Maunder Minimum was one of them, but in these minimum periods the mean cycle was also 11 years. This is consistent with the result of Xu and Jiang.

Tree-ring widths have failed to show convincing evidence of past solar cyclic activity, because local, rather than global effects dominate their patterns.[43] As to C^{14}, it is also impossible to detect the 11-year cycle due to the appreciable delay (10-50 years) between variations in C^{14} production and resultant changes in the biospheric abundance, although Eddy has used C^{14} history to recognize six major excursions in solar behavior in the past 2000 years with possibly a total of 12 in the past 5000 years.[44] Thus, to study solar cycles using natural 'unwritten' records is unsatisfactory, and the solar cycle as well as its origin still remain an unsolved problem in astrophysics.

III. The Gravitational Constant G

The relationship between Ephemeris Time (ET) and Universal Time (UT) is

$$ET = UT + \Delta T \qquad (2)$$

where ΔT is determined by observation. If the steady increase in the length of the day is one millisecond (ms) per century, the mean length of the day over the past 20 centuries is 10 ms shorter than that of the present day, and the sum of ΔT is

$$10 \times 10^{-3} \times 365.25 \times 100 \times 20 = 7305 \text{ s} = 2 \text{ h } 01 \text{ m } 45 \text{ s} .$$

Thus, the difference between the recorded time (UT) of observation of solar eclipses in the first century and the calculated time (ET) according to gravitational theory might be as much as two hours. The earlier a solar eclipse took place, the greater would be the difference in time. Moreover, the difference in time may also be reflected by the difference between the observed place and the calculated path of the eclipse. For example, in the seventh year of Lu-Hou 呂后 (The Queen of Liu Bang 劉邦) Chinese astronomers reported a total eclipse of the Sun at Changan 長安 on a date corresponding to March 4, -181. The calculation of the path of this totality by Stephenson and Clark on the assumption of no de-acceleration of the Earth's spin shows that the track comes out well to the west of Changan. A rotation of the Earth of 75 degrees would be necessary to make the eclipse total at Changan. The inference is that in the intervening time between -181 and the present day, the Earth has lost five hours relative to an ideal clock, i.e., the increase rate of the day length is more than 2 ms/century.[45]

The application of historical records of solar eclipses to studying the de-acceleration of the Earth's rotation can get precisely the same result as current observations do, if the exact date and place of the observation, as well as the magnitude of the eclipse, is described in these records - the greater the magnitude, the better. It would be best if we could obtain historical records of the occurrence of total solar eclipses. In 1980 Wu Shou-Xian 吳守賢 checked Chinese eclipse records used by four Western astronomers (D. R. Curott, 1966; R. R. Newton, 1970; P. M. Mullar and F. R. Stephenson, 1975). He found out that, apart from repetition of each other, the total number of records they used is 30, one-third of which are incorrect.[46] Consequently, their results should be questionable.

Another new time system, Atomic Time (AT), was proposed in 1967, for the intrinsic accuracy of Ephemeris Time is not easy to achieve. It utilizes the natural resonance frequency of Cs^{133} to make extremely precise measurements of time intervals, and is not related to gravitational force. If the universal constant of gravitation, G, varies with time as Paul Dirac predicted, there would be a difference between the results of historical solar eclipses reckoned by Ephemeris Time and by Atomic Time.

In 1975 Van Flandern made an interesting observational test based on modern and historical dates. His analysis of lunar occultations over the period 1955 to 1974 gave a result for G:

$$\frac{1}{G}\left|\frac{dG}{dt}\right| = \dot{G}/G = (-8 \pm 5) \times 10^{-11}/\text{year}\ , \tag{3}$$

i.e., G is on the decrease.[47] But this problem has recently been analyzed again using lunar laser ranging measurements and eclipses data from ancient China. The close agreement between both results is evidence in favor of the constancy of G.[48] However, definite evidence against (or for) the non-constancy of G must await a further study.

In addition to the topics discussed above, there are many others in astrophysics where the historical records may have applications, such as in cometary investigations, planetary orbital perturbation studies, solar-terrestrial relationship research, etc. We believe that combining current technology with information recovered from the records of our ancestors will certainly benefit the study of astrophysics and increase our understanding of the universe. We have collected and arranged all historical records of celestial phenomena in China, and are now preparing to publish them.

References

1. A talk with Wu You-Xun 吳有訓 in 1972 at Beijing 北京.
2. K. Lundmark, Publication of Astronomical Society of the Pacific, 33, 225 (1921).
3. E. Hubble, Ast. Soc. Pacific Leaflet, No. 14, 1928.
4. J. J. Duyvendark and J. H. Oort, T'oung Pao《通報》, (T'oung Pao Archives), 36, 174 (1972).
5. J. J. Duyvendark, Publication of Astronomical Society of the Pacific, 54, 91 (1942)
6. N. U. Mayall and J. H. Oort, ibid., 54, 95 (1942).
7. L. I. Sedov, Similarity and Dimensional Methods in Mechanics (Academic Press, New York, 1959).
8. D. H. Clark and J. L. Caswell, Monthly Notices of Royal Astronomical Society, 174, 267 (1976).
9. B. R. Goldstein and Ho Peng-Yoke 何丙郁, Astronomical Journal, 70, 748 (1965).
10. F. F. Gardner and D. K. Milime, Astronomical Journal, 70, 754 (1965).
11. F. R. Stephenson et al., Monthly Notices of Royal Astronomical Society, 180, 567 (1977).
12. Bo Shu-Ren 薄樹人 and Wang Jian-Min 王健民, Ke-Ji-Shi Wen-Ji《科技史文集》(Collected Essays on Science and

Technology), 1, 79 (1978), a special issue on astronomy.

13. Wang Jian-Min 王健民, Beijing Tian-Wen-Tai Tai-Kan 《北京天文台台刊》(Publications of Beijing Astronomical Observatory), 1, 69 (1979).

14. Fang Li-Zhi 方勵之 and Qu Qin-Yue 曲欽岳, Zhong Guo Ke Xue 《中國科學》(Scientia Sinica), 4, 406 (1978).

15. Xi Ze-Zong 席澤宗, Tian-Wen Xue-Bao 《天文學報》(Acta Astronomica Sinica), 3, 183 (1955).

16. Xi Ze-Zong 席澤宗 and Bo Shu-Ren 薄樹人, Tian-Wen Xue-Bao 《天文學報》(Acta Astronomica Sinica), 13, 1 (1965).

17. G. G. C. Palumbo et al., Tian-Wen Xue-Bao 《天文學報》(Acta Astronomica Sinica), 21, 334 (1980).

18. Li Qi-Bin 李啓斌, Tian Wen Xue-Bao 《天文學報》(Acta Astronomica Sinica), 19, 210 (1978).

19. K. Imaeda and T. Kiang 江濤, Journal for the History of Astronomy, 11, 77 (1980).

20. P. E. Angerhofer, Archaeoastronomy, 4, 22 (1981).

21. Wang Zhen-Ru 汪珍如 and F. D. Seward, Astrophysical Journal, 285, 607 (1984).

22. Wang Zhen-Ru 汪珍如 et al., Highlights of Astronomy, 7 (to be published). Presented at the Supernova Joint Discussion and Working Group Session in the IAU General Assembly, New Delhi, 1985.

23. Wang Zhen-Ru 汪珍如, Invited Talks at IAU Symposium no. 125, 'Origin and Evolution of Neutron Stars," Nanjing 南京, 1986.

24. W. Baade and R. Minkowski, Astrophysical Journal, 119, 206 (1954).

25. Sun-Il Chu, Journal of the Korean Astronomical Society, 1, 29 (1968).

26. S. F. Gull, Monthly Notices of Royal Astronomical Society, 161, 47 (1973).
27. Sidney Van den Bergh and W. W. Dodd, Astrophysical Journal, 162, 485 (1970).
28. W. B. Ashworth, Journal for the History of Astronomy, 11, Part 1 (1980).
29. K. Brecher and I. Wasserman, Astrophysical Journal, 240, 105 (1980).
30. R. P. Broughton, Journal of Royal Astronomical Society of Canada, 173, 381 (1980).
31. K. W. Kamper, Observatory, 100, 3 (1980).
32. Bo Shu-Ren 薄樹人 and Wang Jian Min 王健民, Ke-Ji-Shi Wen-Ji《科技史文集》(Collected Essays on Science and Technology), 1, 75 (1978), a special issue on astronomy.
33. K. Brecher, Bulletin of American Astronomical Society (1980).
34. W. Maunder, Monthly Notices of Royal Astronomical Society, 50, 251 (1890).
35. J. A. Eddy, Science, 192, 1189 (1976); Proceedings of the Inter. Sym. on Solar-Terr. Phys., 11, 958 (1976); Sky and Telescope, 52, 394 (1976).
36. Ding You-Ji 丁有濟 and Zhang Zhu-Wen 張築文, Ke-Xue Tong-Bao《科學通報》(Science Bulletin), 23, 107 (1978).
37. Luo Bao-Rong 羅葆榮 and Li Wei-Bao 李維葆, Ke-Xue Tong-Bao《科學通報》(Science Bulletin), 23, 262 (1978).
38. Zou Yi-Xin 鄒儀新, Beijing Tian-Wen-Tai Tai-Kan《北京天文台台刊》, Publications of Beijing Astronomical Observatory, 12, 87 (1978).
39. Xu Zhen-Tao 徐振韜 and Jiang Yao-Tiao 蔣窈窕, Nanjing Da-Xue Xue-Pao《南京大學學報》(Journal of Nanjing University), 2, 31 (1979).

40. Kanda Shigeru 神田茂, Tokyo Temmondai Ho《東京天文台報》(Publications of Tokyo Astronomical Observatory), 1, No. 1 (1932).

41. J. A. Eddy, Archaeoastronomy, No. 4, 9 (1982).

42. Dai Nian-Zu 戴念祖 and Chen Mei-Dong 陳美東, Ke-Ji-Shi Wen-Ji《科技史文集》(Collected Essays on Science and Technology), 2, 69 (1980).

43. La Marche and H. C. Fritts, Tree Bulletin, 32, 21 (1972).

44. J. A. Eddy, The Solar Output and Its Variation (Colorado Associated Univ. Press, Boulder, 1977).

45. F. R. Stephenson and D. H. Clark, Quarterly Journal of Royal Astronomical Society, 18, 340 (1977).

46. Wu Shou-Xian 吳守賢, Shanxi Tian-Wen-Tai Tai-Kan《陝西天文台台刊》(Publications of Shaanxi Astronomical Observatory), 2, 23 (1980).

47. T. C. Van Flandern, Monthly Notices of Royal Astronomical Society, 170, 333 (1975).

48. F. R. Stephenson, New Scientist, 81, 560 (1979).

49. F. R. Stephenson and D. H. Clark, Applications of Early Astronomical Records (Adam Hilger Ltd., Bristol, 1978).

OPTICS IN CHINA BASED ON THREE ANCIENT BOOKS

Jing-Guang Wang 王錦光

I. The Mo Jing《墨經》

The Mohist school was founded by the great philosopher Mo Zi 墨子 (-478 to -392). The Mo Jing《墨經》(The Mohist Canon) was written between -450 and -250. It treats logic, epistemotogy, geometry, mechanics, and optics. The treatment of optics is given in eight 'ming ti' 命題 ('propositions'). Excellent accounts of Mohist optics in English are now available.[1,2] Here, I will just provide a brief summary.

The first five 'ming ti' are on the theory of shadows.

1. The first 'ming ti' discusses the formation and vanishing of shadows.
2. The second 'ming ti' concerns umbras and penumbras.
3. The third 'ming ti' presents rectilinear propagation of light.

   Light shines on a person as a flying arrow, i.e. the light travels in a straight line. A man standing in sunlight shines as though he were radiating rays. If a pinhole chamber is used, the bottom part of the man becomes the top part of the image whereas rays from the top part become the bottom part of the image. Suppose his foot emits light rays, some rays would strike the place below the pinhole while others would pass through the pinhole to form the top part of the image. On the other hand, his head sends out rays, some of which would strike above the pinhole but others penetrate to form the lower part of the image. The complete image is inverted (Figure 1) due to the intersection ('wu' 午) of rays at a point ('duan' 端).

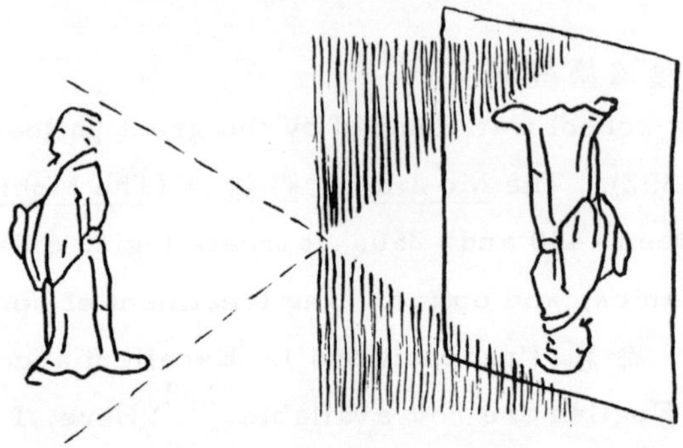

Figure 1. Pinhole image.

This 'ming ti' is a true record of experiments on pinhole images. Its explanation is very clear because it is based upon the rectilinear propagation of light.

4. The fourth 'ming ti' describes the reflection of sunlight and the formation of an inverted shadow of a man.
5. In the fifth 'ming ti,' the phenomenon of changing shadow size is presented: how to make a shadow longer or shorter, narrower or wider.

After shadow theory, image theory is discussed in 'ming ti' six through eight.

6. The sixth 'ming ti' concerns image formation by a plane mirror and includes an explanation of the meaning of symmetry.
7. The seventh 'ming ti' states that images from a concave mirror are of two varieties: erect and inverted.
8. The eighth 'ming ti' states that images from a convex mirror, however, are of only one variety.

These eight 'ming ti' constitute a systematic theory of geometrical optics which not only has experimental results but also presents theory.

II. The Meng-Qi Bi-Tan 《夢溪筆談》

The Meng-Qi Bi-Tan 《夢溪筆談》(Meng-Qi Essays) was written by the scientist Shen Gua 沈括 (1033-1097) of the Song dynasty (960-1279). There are primarily ten sections dealing with optics; two of them will be discussed in detail.

A. Article 44. Shen Gua discusses image formation by pinholes and concave mirrors

The burning mirror ('yang sui' 陽燧) reflects light so as to form inverted images。 This is due to a focal point ('ai' 礙) being between the object and the mirror. Mathematicians call this study 'ge shu' 格術 . It is analogous to rowing where an oar moves against the oarlock. The phenomenon can also be described in the following way: when a bird flying in the sky casts a shadow,

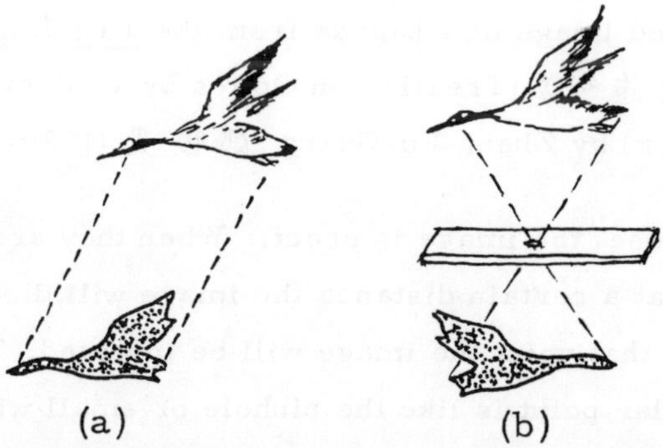

Figure 2(a). Bird and shadow move in the same direction.

(b). Bird and pinhole image move in opposite directions.

its shadow on the ground moves in the same direction as the bird, Figure 2a. However, if its image is formed through a small window (pinhole), the shadow moves in the opposite direction, so that the shadow moves eastward for a bird flying westward, Figure 2b. There is another way to describe it. The image of a pagoda is inverted after being "collected" through a hole or small window (Figure 3). The principle is the same as the burning mirror. Since the surface is concave, the image formed by the mirror varies in size as the distance between the mirror and the object (say a finger) changes. If they are

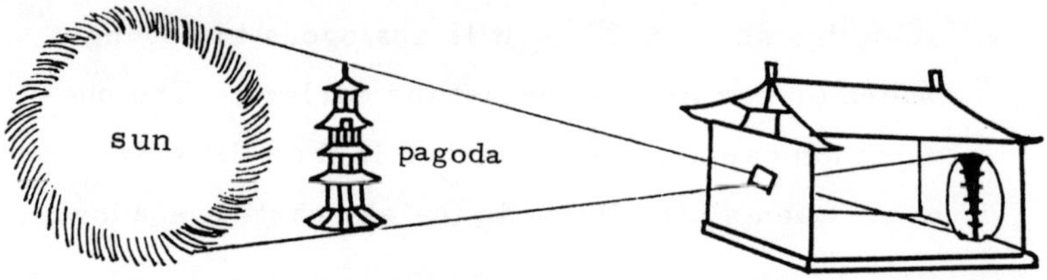

Figure 3. Inverted image of a pagoda from the Jing Jing Ling Chi 《鏡鏡詅癡》 (Treatise on Optics by an Untalented Scholar) by Zheng Fu-Guang 鄭復光 (1780-?).

very close, the image is erect. When they are separated, at a certain distance the image will disappear. Beyond that point the image will be inverted. This particular point is like the pinhole or small window. Similarly, the handle of an oar is always near one end of the oar, which is supported by the oarlock (which acts as a fulcrum) near the middle.

The oarlock constitutes a kind of 'pivot point' ('yao' 腰, literally 'waist'). (A waist drum is shown in Figure 4 to demonstrate the 'pivot point.') Such opposite motion can also be observed as follows: when one's hand moves upward, the pinhole image moves downward and vice versa.

Because of the concavity of a burning mirror, it can collect sunlight and direct it to a tiny spot no bigger than a hempseed one or two inches away from the mirror surface. This is the place at which to ignite a fire. It is the smallest 'pivot point.'

Figure 4. Waist drum from a wall painting in a tomb of the Song dynasty.

<u>Commentary</u>. The above statement contains three important observations.

(A) The pinhole image. The <u>Mo Jing</u> discussed this problem at a much earlier date but could have had little influence because this work had disappeared in the fourth century. Moreover, Shen Gua explained more lucidly than the Mo Jing, the flying bird (Figure 2), for example.

The Meng-Qi Bi-Tan stimulated Shen Gua's successors to further research. Among them was Zhao You-Ging 趙友欽 (Part III).

(B) Ignition by sunlight. Before Shen Gua's time the use of a concave bronze mirror to make fire had been mentioned in several books such as the Zhuang Zi《莊子》(The Book of Master Zhuang) of the -4th century and the Huai Nan Zi《淮南子》(The Book of the Prince of Huai Nan) of the -2nd century, but none mentioned the focal length. In the Meng-Qi Bi-Tan the focal length is given as one or two inches and the focal point as no larger than a hempseed: a precise description.

(C) 'Ge shu.' Image formation using a pinhole or concave mirror is related to the concept of a focal point ('ai' 礙). The mathematical term was 'ge shu,' but this term disappeared at a later date. It was the Meng-Qi Bi-Tan which stimulated later research on this subject. Zheng Fu-Guang 鄭復光 (1780-?) thought 'ge' meant separation, so the pinhole and the focal point are both 'ge.' Zou Bo-Qi 鄒伯奇 (1819-1869) wrote a book concerning geometric optics which he titled Ge Shu Bu《格術補》(Supplement of 'Ge Shu').

B. Article 330. Shen Gua describes the 'light penetrating mirror' ('tou-guang jian' 透光鑒) as follows:

> There are about twenty characters inscribed on the mirror in ancient style. When the sun shines on a light-penetrating mirror, the characters on the back appear in the image reflected on a wall (Figure 5). It is amazing that the characters can 'pass through' the mirror from the back, and be seen on the wall distinctly.

Figure 5. The pattern of light penetrating mirror is displayed on wall.

Experts told Shen that when a bronze mirror is cast, the thinner regions cool faster than the thick portions, representing this design. Minute wrinkles develop during the process so there exist faint traces on the face that cannot be seen by the eye. Shen Gua also states:

> My family has three mirrors like this and I have seen others treasured in my friends' homes. They are antique and similar in that they 'allow light to pass through.' I do not know why other extremely thin mirrors do not let light pass through.

Commentary. In the Western Han dynasty (-260 to +24) of China, a kind of light penetrating mirror existed. If such a mirror faced the sun, the light reflected on the wall would show the same picture as that on the back of the mirror. Since the light penetrated the mirror and was reflected from the back, the ancient Chinese called it a 'light penetrating mirror.' 'Magic mirror' is another name used in the West and Japan.

Today, research in China has found several ways to make a 'magic mirror.' The research benefitted from the Meng-Gi Bi-Tan, since Shen Gua provided the key for the making of a 'magic mirror' with his observation of a pattern having faint lines on the back. Since Shen's time, many scholars have been interested in this mysterious phenomenon. Zheng Fu-Guang 鄭復光 , among them, gave a detailed description and analysis in his Jing Jing Ling Chi 《鏡鏡詅癡》(Treatise on Optics by an Untalented Scholar). He apparently understood 'light penetration.' He pointed out that, in order to create such a mirror with different thicknesses, pressure on the polishing tool should be varied in accordance with the design.

The Meng-Qi Bi-Tan gives, in addition to articles 44 and 330, the quantitative relation between the size of the image and the curvature of a convex mirror. It introduces a method for conducting a postmortem by using red light. (If a person is whipped before death, blood will accumulate under the skin which is invisible under normal illumination. If red light is used the blood and the whipping become evident.) Shen Gua ascribes the rainbow to the sun's shadow during rain. He also discusses the incomplete or broken window, mirages and the fluorescence of living tissues.[3]

III. The Ge Xiang Xin Shu 《革象新書》

The Ge Xiang Xin Shu 《革象新書》(New Astronomy) was written by the Taoist Zhao You-Qin 趙友欽 , who lived in the 13th century.[4] One chapter in the book, 'Pinhole Images,' is very valuable for its introduces a giant experiment conducted in a two story house. Two circular wells four feet in diameter were dug in two downstairs rooms. The depth of the right well was 4 feet, of the left, 8 feet. A 4 foot tall table could be put into the left well if needed (Figure 6). Two light sources were prepared in the same manner: more than one thousand candles were inserted into a board four feet in

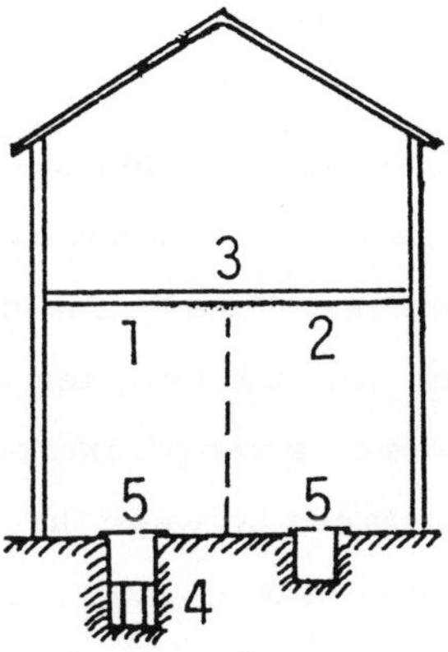

1. Left room with a well.
2. Right room with a well.
3. Ceiling of the first floor used as the fixed screen.
4. Table.
5. Well-cover with a hole in the center.

Figure 6. An illustration of Zha You-Qin's 趙友欽 experiment of the 13th century.

diameter. After the light source board was placed on the bottom of the well (or on the table), each well was covered by another board with a hole in the center. The ceiling of the first floor was used as the fixed image screen and an adjustable screen was suspended from the ceiling.

Such an experimental arrangement is very reasonable. First, the flame is relatively stable because the burning candles are in the well. Second, if light can pass only through the hole in the cover of the well, observation is easy and the results accurate. Third, since the wells are quite deep, the distance between the source and screen can be varied over a wide range. Fourth, comparison can be made between the left and right rooms under various conditions: changing object distances, varying image distances, etc. Therefore the experiment can be performed using a single variable technique, changing one variable while others remain fixed. The light source possesses the property of wide extension, its shape and strength can be adjusted by lighting the proper candles. The maximum luminosity can be as high as one thounsand candles if the luminosity of each candle is unity, so the experiment can be carried out during the daytime. Such a laboratory shows great sophistication.

Zhao obtained the following results after a series of experiments: (1) the shape of the pinhole image is independent of the shape and size of the small hole; (2) the brightness of the image depends on the size of the small hole, the larger the hole the brighter the image; (3) the brightness increases when the source strength increases; (4) the brightness decreases when the distance between the source and image increases. (In the West, four centuries later, the German scientist J. H. Lambert (1728-1777), obtained the quantitative relation: brightness is inversely proportional to the square of the distance.)

It is worthy of notice that Zhao not only searched experimentally for regularities of image formation and brightness but he also gave theoretical arguments. His basic idea was: (1) there exists a light spot on the screen corresponding to a single candle; (2) if one thousand candles are burning there should exist one thousand images. These images may overlap. The whole image changes as the spacing of the candles change. It is evident that Zhao understood the principle of the rectilinear propagation and superposition of light. Zhao also used the pinhole image to study eclipses of the sun and the moon as well as other astronomical phenomena.

Zhao's giant experiment is more than three centuries earlier than the 'revolutionary' experiments of Galileo (1564-1642).

Optics in ancient China as attested by the Mo Jing, the Meng-Qi Bi-Tan and the Ge-Xiang Xin-Shu not only began as early as the optics in Greece but also attained a high level of achievement both in theory and in practice.[5]

References

1. Joseph Needham, Science and Civilization in China (University Press, Cambridge, 1962), Vol. 4, Part I, pp. 78-125.
2. A. C. Graham and Nathan Sivin, 'A Systematic Approach to the Mohist Optics' in Chinese Science (The MIT Press, Cambridge, 1973), pp. 105-152.
3. Wang Jing-Quan 王錦光 and Wen Ren-Jun 聞人軍, 'Shen Gua de Ke-Zue Cheng-Jiu Yu Gong-Xian'〈沈括的科學成就與貢獻〉('The Scientific Achievements of Shen Gua') in Shen-Gua Yan-Jiu《沈括研究》(Research on Shen Gua's Work) (Zhejiang 浙江 People's Publisher, 1985), pp. 64-123.
4. Wang Jing-Quan 王錦光, 'Zhao You-Qin Ji Qi Guang-Xue Yan-Jiu'〈趙友欽及其光學研究〉('Zhao You-Qin and His Research in Optics') in Journal of the History of Science Technology, 12, 93-99 (1984).
5. Wang Jing-Quan 王錦光 and Hong Zhen-Huan 洪震寰, Zhong-Guo Guang-Zue Shi《中國光學史》(History of Optics in China) (Hunan 湖南 Education Publisher, 1986).

THE GENERATION OF CHROMATIC SCALES IN THE CHINESE BRONZE SET-BELLS OF THE -5TH CENTURY*

Cheng-Yih Chen 程貞一

I. Introduction

In March 1978, 124 musical instruments were unearthed from the tomb of Hou Yi 侯乙 (Marquis Yi) of the state Tseng 曾 in the present Sui 隨 county, Hupei 湖北 province.[1] Among these instruments, there are string instruments such as the 'se' 瑟 (zither) and 'ch'in' 琴 (half-tube zither), wind instruments such as the 'sheng' 笙 (reed mouth-organ) and 'p'ai-hsiao' 排簫 (pan-pipes), and percussion instruments such as the 'ku' 鼓 (drums) and 'pien ch'ing' 編磬 (stone set-chimes).[2] But most magnificent is a set of bronze 'pien chung' 編鐘 (set-bells), 64 in total (see Figure 1). Textual information about the instruments and tonal systems is found in the inscriptions on both the bronze set-bells and stone set-chimes and in the explanatory writings on their racks and containers.[3-6]

From the inscriptions found on the 'po chung' 鎛鐘 (ceremonial bell), a gift to Marquis Yi from King Chu 楚 (the fifth bell from the right on the lower rack in Figure 1), archeologists have determined that the tomb was sealed around -433.[1] This allows us to date the set-bells to the middle of the -5th century. In fact, the set-bells were not all cast in the same period. The 45 'yung chung' 甬鍾

---

*A preliminary report of this work, entitled 'Intervallic Structure of Unearthed Bronze Set-Bells of the -5th Century' was presented at the XVIIth International Congress of History of Science, University of California, Berkeley, July 31-August 8, 1985

Figure 1. A photograph of the bronze set-bells ('pien chung' 編鐘) unearthed in 1978 from the tomb of Marquis Yi 乙 of the state Tseng 曾 in the present Sui 隨 county. Hupei 湖北 province. The largest bell weighs 203.6 kg with a height of 153.4 cm and the smallest weighs 2.4 kg with a height of 20.4 cm.

(bells with stems) located in the middle and lower racks in Figure 1 form a chromatic tonal system[7, 8] and were constructed in accordance with the specification of Marquis Yi around -450, since engraved on the largest and heaviest 'yung chung' (the first bell from the left on the lower rack in Figure 1) is the statement: "Marquis Yi, the maker and the owner."[1] The remaining 19 set-bells are all 'niu chung' 鈕鐘 (bells with knobs) located on the top rack.

There are indications that these 'niu chung' were originally hung in the right section of the middle rack. On the crossbar of the right middle rack, a row of holes made to fit the knobs of the 'niu chung' for hanging may still be found. The top rack was probably added for the 'niu chung' to make room for the 'yung chung' cast during the year when Marquis Yi was enthroned at an age of around 25 years or shortly after. It has been suggested that these 'niu chung' were used as pitch standards.[7, 8] T'an Wei-Szu 譚維四 and Feng Kuang-Sheng 馮光生 have suggested that parts of the 'niu chung' also form a chromatic tonal system.[9] There is no doubt that the 'niu chung' were cast in an earlier period than the 'yung chung.'

In this paper, the acoustic properties and intervallic structure of the 'niu chung' set-bells are examined. By the use of the measured frequencies for the set-bells and the textual information provided by the inscriptions, the mechanism used by the early Tseng 曾 musicians for the generation of chromatic scales is deduced. And a comparison of this mechanism with that described in the Lü-Shih Ch'un-Ch'iu 《呂氏春秋》 (Master Lü's Spring-Autumn Annals) of -239 is then made.

II. Acoustic Analysis of the Niu Chung 鈕鍾

A. Frequencies and Modes of Vibration

The bronze set-bells, when discovered, were submerged under water in one of the chambers of the tomb.[1] This was fortunate since the water, with its low oxygen adsorption, protected the bronze from substantial corrosion. As a result, the bells are well-preserved.

Each of the bells is marked with two strike points and was undoubtedly designed to produce two modes of vibration with different frequencies. Since a bell with a perfectly symmetrical mass distribution can exhibit only one mode of vibration, all the set-bells are necessarily asymmetrical in their mass distribution. The use of asymmetry in mass distribution to obtain, in each bell, an extra mode of vibration, is certainly inventive and efficient but it requires rather advanced acoustic analysis so that each mode can be individually excited without appreciable interference. We now know that only when the nodal lines of one of the vibrational patterns fall along the antinodal lines of the other can the two modes of vibration of a particular bell be individually excited to produce their corresponding resonance frequencies without interference.

The frequencies produced by exciting each mode of vibration at the designated strike point have been measured for all of the set-bells by the Peking 北京 Music Institute using a stroboscope.[7] Independently, these frequencies were also measured using a tonometer by P'an Tu-Wu 潘篤武 and Liu Kuei-Hsing 劉貴興 of the physics department of the Fu Tan 復旦 University and by Ma Ch'eng-Yüan 馬承源 and P'an Chien-Ming 潘建明 of the bronze division of the Shanghai 上海 Museum.[10] The two sets of measurements are reproduced in Table I for the 19 'niu chung' 鈕鐘 set-bells located on the upper rack. The bells are identified by the code, Top n-m (i.e. the mth bell in the nth section of the top rack) assigned when

Table I. Measured Frequencies of Set-Bells in the Upper Rack

	Front Strike-Point Mode					Side Strike-Point Mode				
Assigned Bell Code[a]	Note Name 階名	Measurements A[b] vibs/s	Measurements A[b] cents	Measurements B[c] cents	ΔC	Note Name 階名	Measurements A[b] vibs/s	Measurements A[b] cents	Measurements B[c] cents	ΔC
Top 1-1	羽曾	1990	8312	8305	7	宮曾	2339	8592	8585	7
Top 1-2	徵角	1393.4	7695	7688	7	徵曾	1784.2	8123	8110[d]	13
Top 1-3	商角	1100	7286[e]	7280	6	商曾	1377.4	7675	7670	5
Top 1-4	徵曾	731	6578	6575	3	徵	918.2	6973	6965	8
Top 1-5	羽角	651	6378	6374	4	羽曾	812	6761	6755	6
Top 1-6	羽	520	5989	5984	5	宮	619	6291	6281	10
Top 2-1	商曾	1390	7691	7681	10	羽角	1678.7	8018	8008	10
Top 2-2	商角	1101	7288	7278	10	羽	1306.8	7584	7576	8
Top 2-3	商	822.7	6783	6775	8	羽曾	998	7117	7115[f]	2
Top 2-4	商曾	683.5	6462	6457	5	羽角	818.9	6775	6767	8
Top 2-5	商角	543.5	6065	6058	7	羽	642.6	6355	6350	5
Top 2-6	商	410.1	5578	5568	10	羽曾	500.5	5923	5912	11
Top 3-1	商	1781	8120	8115	5	羽曾	2117	8419	8415	4
Top 3-2	宮曾	1243	7503	7490	13	徵角	1447	7761	7755	6
Top 3-3	宮角	944	7021	7011	10	徵	1140	7348	7340	8
Top 3-4	宮	731.7	6580[g]	6573	7	徵曾	890	6919	6911	8
Top 3-5	宮曾	599.3	6235	6235	0	徵角	716.4	6543	6537	6
Top 3-6	宮角	471.9	5821	5811[h]	10	徵	571.2	6152	6145	7
Top 3-7	宮	365.1	5377	5363	14	徵曾	446.6	5725	5715	10

[a] The bells are labeled by the code, Top n-m (i.e. the mth bell in the nth section of the top rack) assigned when they were first unearthed.

[b] Measurements made jointly by the physics department of Fu-Tan University and the bronze division of the Shanghai Museum (see Ref. 10).

[c] Measurements made by the Peking Music Institute (see Ref. 7).

[d] The note A_6 - 10 cents in Ref. 7 should read A_6 + 10 cents.

[e] The entry 7206 cents for 1100 vibs/s in Ref. 10 should read 7286 cents.

[f] The note B_5 - 15 cents in Ref. 7 should read B_5 +15 cents.

[g] The entry 6700 cents for 731.7 vibs/s in Ref. 10 should read 6580 cents.

[h] The note $B^{\#}_4$ - 89 cents in Ref. 7 should read B_4 - 89 cents.

they were first unearthed. It is seen from the table that the measured frequencies for each mode of vibration are in excellent agreement. The results indicate that these bells are capable of generating their characteristic modes of vibration with reproducible frequencies when these modes are excited at the designated strike points.

The fact that the two notes in each bell can be separately produced without appreciable beats suggests strongly that the bell-makers were rather successful in superimposing the antinodal lines of one mode on the nodal lines of the other. Thus, the front strike points locates the position at which an antinode of the lower vibrational mode is superimposed on a node of higher vibrational mode. The side strike point locates the position of the reverse. This remarkable achievement attests to not only the bell-makers' mastery in campanology but also their advanced knowledge of acoustics.

B. Intervallic Structure

The nineteen 'niu-chung' are not all of the same tonal system. The right three bells (Top 1-4, Top 1-5 and Top 1-6) form a separate tonal system from the bells in sections 2 and 3. The left three bells in section 1 (Top 1-1, Top 1-2 and Top 1-3) do not form a tonal system and are not compatible with the right three bells in section 1, but are compatible with the bells in sections 2 and 3.

The relationship between the left three bells of section 1 and the bells in sections 2 and 3 is summarized in Table II. Except for the 'kung tseng' 宮曾 note of the Top 1-1 bell with the highest frequency (2339 vibs/s), all notes of the left three bells are compatible with the tonal system made up of the notes of the bells of sections 2 and 3. This suggests that these three bells were probably used as guides for standard pitches in the construction of the set-bells in sections 1 and 2.

Table II. The Relationship between the Left Three Bells in Section 1 and the Bells in Sections 2 and 3.

Note Names	Frequencies vibs/s	Frequencies vibs/s	Relationship
'shang-chio' 商角	1100 (Top 1-3)	1101 (Top 2-2)	Same note
'shang-tseng' 商曾	1377.4 (Top 1-3)	683.5 (Top 2-4)	Octave
'chih-chio' 徵角	1393.4 (Top 1-2)	716.4 (Top 3-5)	Octave
'chih-tseng' 徵曾	1784.2 (Top 1-2)	890 (Top 3-4)	Octave
'yü-tseng' 羽曾	1990 (Top 1-1)	998 (Top 2-3)	Octave
'kung-tseng' 宮曾	2339 (Top 1-1)	1243 (Top 3-2)	-

The intervallic structure of the notes of the right three bells in section 1 (Top 1-4, Top 1-5 and Top 1-6) is summarized in Table III. For comparison, we include in Table III the intervals of the

Table III. Intervallic Structure of the Notes of Three Set-Bells (Top 1-4, Top 1-5 and Top 1-6).

	Measured Frequency[a]		Intervals in Cents		
Note Names	vibs/s	cents	Tseng Intonation	Natural Intonation	Tempered Intonation
'yü' 羽	520	5989			
			301	315.6	300
'kung' 宮	619	6291			
			87	70.7	100
'yü-chio' 羽角	651	6378			
			200	203.9	200
'chih-tseng' 徵曾	731	6578			
			183	223.5	200
'yu-tseng' 羽曾	812	6761			
			212	203.91	200
'chih' 徵	918.2	6973			

[a]Based on measurements given in Ref. 10.

corresponding notes in natural (just) intonation and the equal-temperament intonation. The Tseng intervals are in reasonable agreement with those of both the natural and equal-temperament intonations. The six notes of these three bells possibly constitute an early example of an hexatonic scale since it is known from later writings that early Chinese musicians often viewed the chromatic scale as composed of 'yin' 陰 and 'yang' 陽 two complementary hexatonic scales.[11]

The measured frequencies for the notes of the thirteen bells in sections 2 and 3 reveal that these notes make up a chromatic scale as shown in Table IV. Each of these bells is capable of producing a pair of notes. In Figure 2, we illustrate the pattern for how such pairs of

Table IV. The Frequencies of the Notes of a Chromatic Scale Found in the Thirteen Set-Bells of Sections 2 and 3 of the Top Rack (see Figure 1).[a]

Chromatic Sequence	Note Names	1st Octave		2nd Octave		3rd Octave	
		vibs/s	cents	vibs/s	cents	vibs/s	cents
C	'kung' 宮	365.1	5377	731.7	6580		
C#	'yü-chio' 羽角	-	-	818.9	6775	1678.7	8018
D	'shang' 商	410.1	5578	822.7	6783	1781	8120
D#	'chih-tseng' 徵曾	446.6	5725	890	6919		
E	'kung-chio' 宮角	471.9	5821	944	7021		
F	'yu-tseng' 羽曾	500.5	5923	998	7117	2117	8419
F#	'shang-chio' 商角	543.5	6065	1101	7288		
G	'chih' 徵	571.2	6152	1140	7348		
G#	'kung-tseng' 宮曾	599.3	6235	1243	7503		
A	'yü' 羽	642.6	6355	1306.8	7584		
A#	'shang-tseng' 商曾	683.5	6462	1390	7691		
B	'chih-chio' 徵角	716.4	6543	1447	7761		

[a]Based on measurements given in Ref. 10.

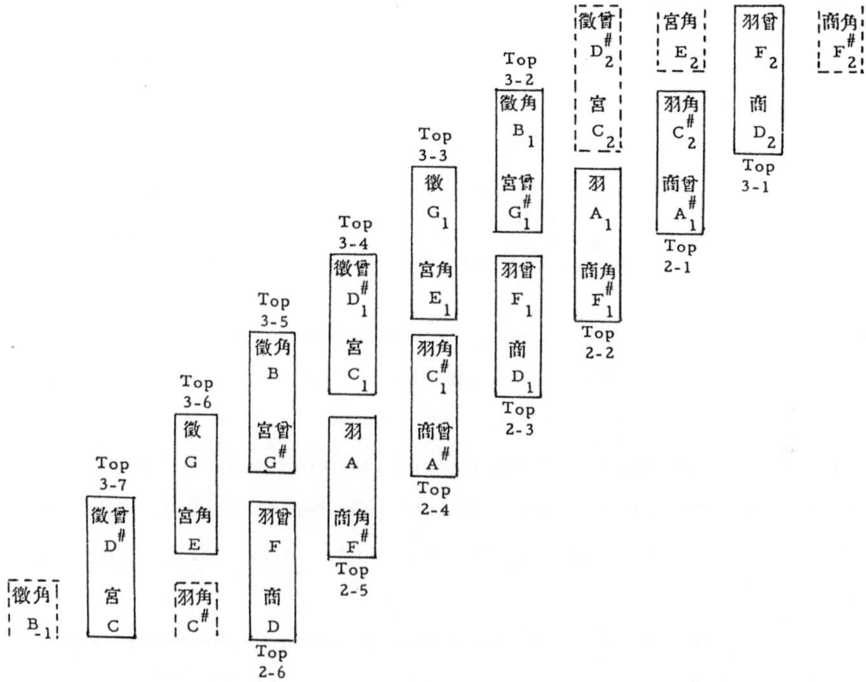

Figure 2. The pattern of a chromatic scale composed of pairs of notes from the set-bells of the -5th century. The names of the notes are found on the bells. The lowest note 'kung' 宮, is arbitrarily identified with the note C to make use of the established chromatic sequence. The bells are labelled by the code, Top n-m (i.e. the mth bell in the nth section of the top rack).

notes make up the chromatic scale in terms of the bells. The identification of the notes with their corresponding bells is made through their code, Top n-m, defined earlier.

For convenience, we arbitrarily identify the lowest note 'kung' 宮, with C to make use of the established chromatic sequence:

$$C, C^{\#}, D, D^{\#}, E, F, F^{\#}, G, G^{\#}, A, A^{\#}, B, C_1, \ldots$$

to illustrate the chromatic relations among the notes. These notes are arranged in asecnding order in Figure 2 from left to right and from bottom to top. It is understood, of course, that these notes are not tuned in accordance with the current convention of A - 440 Hz and their intervals are not tempered as in equal temperament. As shown in Figure 2, the bells from the second and third sections of the top rack, encompass a little over two octaves. Based on the pattern, it is not difficult to see that the bell corresponding to the notes 'kung' 宮 and 'chih tseng' 徵曾, next to the bell denoted by Top 3-2, is, in fact, missing. The bell denoted by Top 3-1, which is found in the position of the missing bell belongs to the second section. The absence of this bell was also predicted by T'an Wei-Ssu 譚維四 and Feng Kuang-Shen 馮光生 based on the pattern in the weight increases of the bells and their relation to the scale.[9]

The systematic manner in which the notes of each bell fit together to form a scale indicates that the bell-makers must have had in their possession a detailed knowledge of the chromatic scale and its intervallic structure prior to the construction of the bells. How, then, was this knowledge obtained? Was this knowledge expressed in terms of certain basic principles? How were the ratios between the notes in the scale determined? Some of the answers to these questions can be deduced from the inscriptions on the bells and from the measured frequencies.

III. Methods for the Generation of Tonal Systems

In this section, the method for generating the chromatic scale found among the 'nui chung' set-bells of Figure 2 will be investigated. To proceed with this investigation, the early Chinese method for generating pentatonic scales is examined in section IIIA. In section IIIB, the method of the Tseng 曾 musicians for the generation of a chromatic scale is investigated and its relationship to the early method for generating pentatonic scales is examined. Finally, in section IIIC, a comparison between the Tseng and Chou mechanisms for the generation of chromatic scales is given.

A. Pentatonic Scales and the Up-and-Down Principle

The oldest extant source of a quantitative procedure for the generation of harmonic progression in musical scales is found in the Kuan Tzu 《管子》 (The Book of Master Kuan). For some reason, this reference on musical scales is not well-known to Western historians of music and science. It is also not cited in Needham's section on acoustics in his monumental treatise Science and Civilisation in China.[12]

The relevant passage on musical scales is found in Chapter 58, entitled 'Ti Yüan' 〈地員〉 ('Elements of Earth'):

凡將起五音，先主一而三之，四開以合九九。

以是生黃鍾小素之首，以成宮；

三分而益之以一，爲百有八，爲徵；

不無有三而去其乘，適足以是生商；

有三分而復於其所，以是生羽；

有三分而去其乘，以是生角。

> As for the generation of the five tones [by a mathematical procedure], one begins first by taking the principal ('chu' 主) (i.e. the agent that produces the tones) to be unity and then multiplying it by the number 3. The fourth power of this number gives 9×9 (i.e. 81) which is the smallest leading number ('hsiao shu chih shou' 小數[素]之首) for the generation of [the tone] Huang-chung 黃鐘 'kung' 宮.
>
> By increasing the number by one third of its value, one obtains the number 108 for [the tone] 'chih' 徵. By decreasing the number so obtained (i.e. 108) by nothing more than one third of its value, one obtains [the number 72] which is sufficient to generate [the tone] 'shang' 商. By giving back to [the number 72] one third of its value, one finds [the number 96] for the generation of [the tone] yü 羽. By decreasing the number so obtained (i.e. 96) by one third of its value, one finds [the number 64] for the generation of [the tone] 'chio' 角.

In current mathematical notation, the procedure can be summarized as follows:

$(1 \times 3)^4 = 9 \times 9 = 81$ for the note 'kung' 宮

$81 \left(1 + \frac{1}{3}\right) = 108$ for the note 'chih' 徵

$108 \left(1 - \frac{1}{3}\right) = 72$ for the note 'shang' 商

$72 \left(1 + \frac{1}{3}\right) = 96$ for the note 'yü' 羽

$96 \left(1 - \frac{1}{3}\right) = 64$ for the note 'chio' 角

It shows that, from an initial number, a sequence of numbers can be generated by consecutively multiplying each generated number by the factors of 4/3 (i.e. 1 + 1/3) and 2/3 (i.e. 1 - 1/3) in alternation, as illustrated in Figure 3.

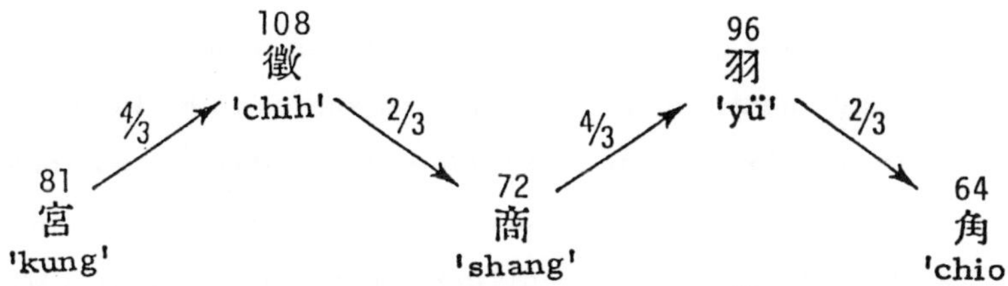

Figure 3. The procedure for generating the pentatonic scale in the 'chih' 徵 mode by the up-and-down principle.

This unambiguously demonstrates that the passage describes a numerical procedure for the generation of a pentatonic scale in the 'chich' 徵 mode with its notes denoted by a set of monosyllabic characters — 'chih' 徵, 'yü' 羽, 'kung' 宮, 'shang' 商 and 'chio' 角, in ascending order. The ratios of the numbers provide the intervallic structure of the scale.

The pentatonic scale in the 'chih' mode is composed of the following harmonic progression:

'chih' 徵	1	fundamental
'yü' 羽	8/9	major second
'kung' 宮	3/4	perfect fourth
'shang' 商	2/3	perfect fifth
'chio' 角	$\frac{16}{27}\left(\sim\frac{3}{5}\right)$	major sixth

Except for the major sixth, the ratios are exactly those found in natural (or just) intonation. The use of a set of monosyllabic characters to designate the degrees of the scale is a type of solmization similar to the movable doh system currently in use.

That the numbers associated with the notes of a scale are relative appears to have been well understood by the early Chinese musicians, since the passage begins with the assignment of unity for the sound producing agent. This unity is then scaled to 81 in the form 9×9 to avoid fractions. The two 9 factors are introduced apparently to account for the two up and two down generations since for each generation the number has to be divided by 3. If the above ratios for the harmonic progressions are reduced to the lowest terms possible on a common base, one obtains the following series of numbers to represent the scale

48	:	54	:	64	:	72	:	81
'chih'		'yü'		'kung'		'shang'		'chio'
sol		la		do		re		mi

This explains why the number 81 is called in the passage 'the smallest leading number.'

The procedure for the generation of the pentatonic scale from the Kuan Tzu is based on physical observations of the acoustics of harmonic sound. Incorporated in the up-and-down generation principle are three harmonic ratios, name the inverse of perfect fourth 4/3, the perfect fifth 2/3 and the perfect octave 1/2 as shown in Figure 4. All of these ratios are exact. As mentioned earlier, the passage on the musical scale from the Kuan Tzu is not well-known among Western historians of music. The incorrect claim that there is no perfect fourth in ancient Chinese music scale continues to appear in some Western accounts.

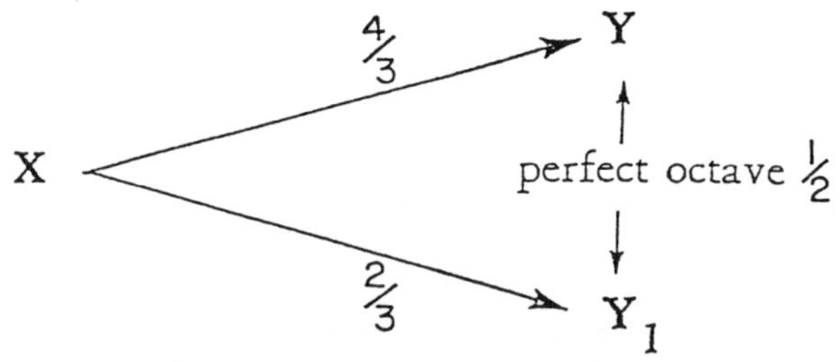

Figure 4. The up-and-down generation principle.

What is of interest here is whether the Tseng musicians of the -5th century knew about the up-and-down principle and the method described in the Kuan Tzu. It is well known that the Kuan Tzu was attributed to Kuan Chung 管仲 (d. -645) in recognition of his contributions as the great statesman of the State Ch'i 齊 but the book was compiled mainly in the -4th century by the academians of the Chi-Hsia 稷下 Academy located in the Ch'i capital. This places the completion of the book a century later than the time of the Tseng musicians. However, one must also take into consideration the evidence indicating that much of the materials compiled in the Kuan Tzu came from early periods of Ch'i. This includes the material in Chapter 58.[13] For a further investigation of this question, it is instructive to examine the intervallic relations among the pentatonic notes found in the 'niu-chung' set-bells.

For the set-bells, we have the measured frequencies for all of their notes. Hence, it is convenient to work with the frequencies directly. In this case, the multiplicative factors, 4/3 and 2/3, in the generating principle are to be replaced by their inverses, namely 3/4 and 3/2, respectively. In Figure 5, the pentatonic scale constructed in accordance with the up-and-down principle is given. Note that in Figure 5 the pentatonic scale is in the 'kung' 宮 mode and is

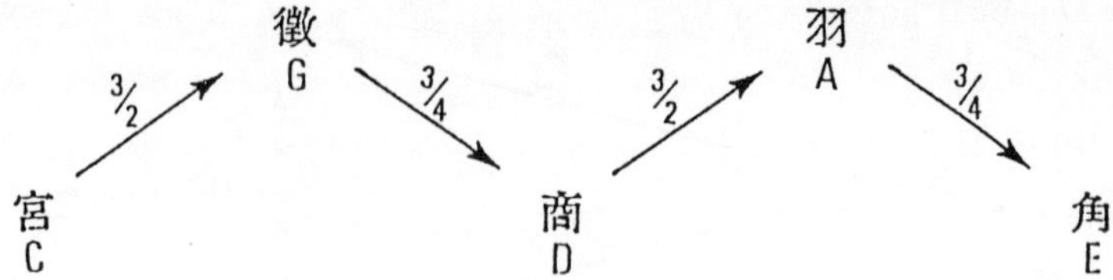

Figure 5. The pentatonic scale generated in accordance with the up-and-down generating principle. The ratios 3/2 and 3/4 are the multiplicative factors for the frequencies of the notes.

given in the reverse sense from that in Figure 3 with the higher notes on top.

A comparison of the ratios of the measured frequencies for the pentatonic notes in the set-bells (see Table IV) with those from the up-and-down principle is given in Table V. It is seen that the ratios

Table V. A Comparison of the Measured Frequency Ratios of the Pentatonic Notes of the 'Niu Chung' Set- Bells with Those by the Up-and-Down Principle.

Frequency Ratio	Measured Ratios		Up-and-Down Principle
	1st Octave	2nd Octave	
'chih' 徵 /'kung' 宫	1.56	1.56	1.5
'shang' 商 /'chih' 徵	0.72	0.72	0.75
'yü' 羽 /'shang' 商	1.57	1.59	1.5
'kung-chio' 宫角 [a]/'yü' 羽	0.73	0.72	0.75

[a]In the Tseng nomenclature, the note 'chio' 角 is called 'kung chio' 宫角 (for an explanation see text below).

of the set-bell notes are in somewhat good agreement with the ratios used in the up-and-down principle, even though the intervals between notes 'chih' and 'kung' and notes 'yü' and 'shang' are slightly larger than the perfect fifth, and the intervals between notes 'chih' and 'shang' and notes 'yü' and 'kung chio' are also slightly larger than the perfect fourth. Based on this comparison, one may conclude that the Tseng musicians were familiar with the intervallic structure of pentatonic scales, and that the procedure described in the Kuan Tzu was probably known also to the Tseng musicians.

B. The Tseng 曾 Chromatic Generation Mechanism

The first clue that is of help in the reconstruction of the procedure by which the Tseng musicians derived their chromatic scale is provided by the names of the notes engraved on the bells. Among the notes of the 'niu chung' set bells under study, there are notes with familiar one-character names such as 'kung' 宮 and 'shang' 商 and unfamiliar two-character names, such as 'chih chio' 徵角 and 'kung tseng' 宮曾. The one-character notes are pentatonic notes satisfying the pentatonic intervallic structure as shown in Table V. The two-character notes are not the usual pentatonic notes but the first character in each of their names is the name of a pentatonic note. One possible interpretation of this is that the two-character notes are generated from the pentatonic notes. This would imply that the Tseng chromatic scale was generated from the pentatonic scale.

It is not difficult to see that the first four notes in the pentatonic configuration are each associated with two additional notes bearing their names as follows:

宮 ⟶ 宮角 ⟶ 宮曾

徵 ⟶ 徵角 ⟶ 徵曾

商 ⟶ 商角 ⟶ 商曾

羽 ⟶ 羽角 ⟶ 羽曾

This suggests that the first four pentatonic notes function as the core from which the additional notes are generated. The notes obtained from the first generation are labelled with the additional character 'chio' 角, while the notes obtained from the second generation are labelled with the additional character 'tseng' 曾. When these eight additional notes are superimposed upon the pentatonic configuration given in Figure 5, one obtains the entire chromatic configuration given in Figure 6 with an extra note 'chio' 角, corresponding to the

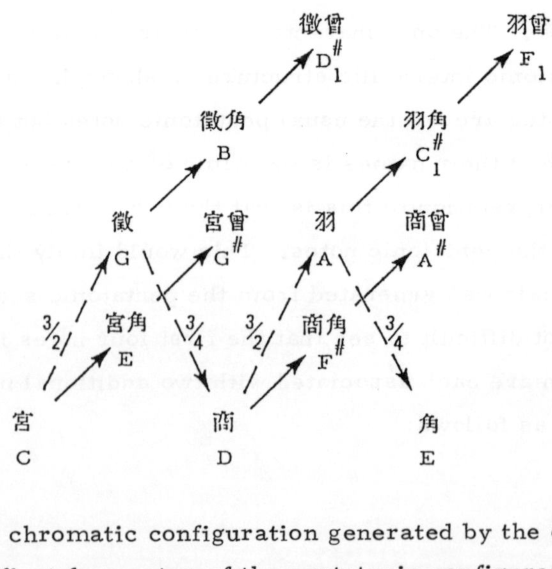

Figure 6. The chromatic configuration generated by the extension of the first four notes of the pentatonic configuration.

note E in the chromatic sequence. But among the notes of the 'niu-chung' set-bells, no note with the name 'chio' is found and in place of it we always find the note 'kung chio' 宫角. This implies that in the Tseng nomenclature the note 'chio' was actually denoted by 'kung chio.'

The twelve notes that make up the chromatic configuration so generated do not all lie within the same octave. By linking such chromatic configurations, however, the complete chromatic scale may be obtained as shown in Figure 7. This can be done by the use of the up-and-down generation principle or simply using the ratio 1/2 for octaves.

The chromatic scale given in Figure 7 is cast in a pattern which is easily identified with the pattern showing how the bells are related to the scale given in Figure 2. The arrows indicate the generating operation. The thirteen bells with a total of twenty-six notes encompass a little over two octaves. Within this range, the lowest note 'yü' chio' 羽角, corresponding to $C^{\#}$ in the chromatic sequence, and the highest note 'kung' 宫, corresponding to the C_2 in the chromatic sequence, are absent (see also Table IV).

The remaining problem, then, is to determine the ratios used in generating the chromatic scale from the pentatonic notes. Since the note 'kung chio' 宫角 is equivalent to the note 'chio' and is generated from its 'parent' note 'kung' 宫, the interval between 'kung' and 'kung chio' is simply that between 'kung' and 'chio.' From Figure 5, it is seen that the note 'chio' is the last note to be generated by the up-and-down principle. Thus, the interval ratio between the notes 'kung' and 'chio' is given by the product $3/2 \times 3/4 \times 3/2 \times 3/4$, i.e. 81/64, which, to a good approximation, is equal to the major third interval 5/4. It then follows that the note 'kung chio' is generated from the note 'kung' with an interval of 'chio' 角 probably given

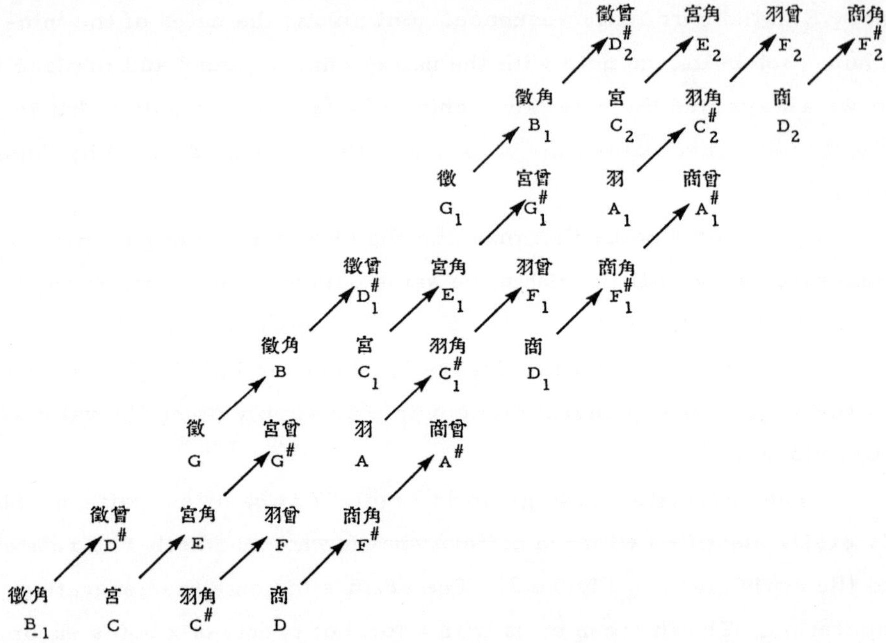

Figure 7. The chromatic scale generated from pentatonic notes in a core configuration. The notes in the scale are arranged in ascending order from left to right and bottom to top matching the order given in Figure 2.

by the ratio 81/64. Similarly, notes 'chih chio' 徵角, 'shang chio' 商角 and 'yü chio' 羽角 are generated from their respective 'parent' notes 'chih' 徵, 'shang' 商 and 'yü' 羽 with probably also the same interval ratio. The interpretation that the character 'chio' 角 specifies the interval for the first generation leads to the expectation that the character 'tseng' 曾 must likewise specify the interval for the second generation.

In Table VI, the ratios of the measure frequencies of the generated notes to their 'parent' notes are tabulated. It is seen that the

Table VI. A Study of the Measured Frequency Ratios between the Pentatonic Notes and the Notes Generated from Them.

Frequency Ratio		Measured Ratios		Theoretical Ratios	
		1st Octave	2nd Octave	Up-and -down	Natural (Just)
'kung chio' 宮角	/'kung' 宮	1.29	1.29	1.27[a]	1.25[b]
'chih chio' 徵角	/'chih' 徵	1.25	1.27	1.27[a]	1.25[b]
'shang chio' 商角	/'shang' 商	1.33	1.34	1.27[a]	1.25[b]
'yü chio' 羽角	/'yü' 羽	1.27	1.28	1.27[a]	1.25[b]
'kung tseng' 宮曾	/'kung' 宮	1.64	1.70	1.60[c]	1.6[d]
'chih tseng' 徵曾	/'chih' 徵	1.56	-	1.60[c]	1.6[d]
'shang tseng' 商曾	/'shang' 商	1.67	1.69	1.60[c]	1.6[d]
'yü tseng' 羽曾	/'yü' 羽	1.55	1.62	1.60[c]	1.6[d]

[a] $81/64 \cong 1.27$.
[b] 5/4 the major third interval.
[c] $(81/64)^2 \cong 1.60$.
[d] 8/5 the minor sixth interval.

ratios for notes of the first generation compare reasonably well with the major third interval 5/4 and its approximation 81/64, and the ratios for notes of the second generation with the minor sixth interval 8/5 and its approximation $(81/64) \times (81/64)$. The exceptions are the ratios between the notes 'shang chio' and 'shang' which approximates the perfect fourth interval 4/3 and ratios between the notes 'shang tseng' and 'shang' and the note 'kung tseng' and 'kung' which approximate the major sixth interval 5/3. In view of the age of these bells and the technical complexities in their construction and tuning, a definite conclusion cannot be made on the ratios simply based on the measured frequencies. The comparison does support the interpretation that the characters 'chio' 角 and 'tseng' 曾 are indicative of the intervals in the first and second generation, respectively.

C. A Comparison with the Chou 周 Chromatic Generation Mechanism

The Tseng mechanism for the generation of a chromatic scale can be compared with the Chou 周 mechanism since a description of the latter is found in the Lü-Shih Ch'un-Ch'iu 《呂氏春秋》 (Master Lü's Spring-Autumn Annals) of -239. We have from Chapter 6 'Yin Lü' 〈音律〉 ('Tonal Pitch') the passage:

黃鐘生林鐘，林鐘生太蔟，太蔟生南呂，南呂生姑洗，姑洗生應鐘，應鐘生蕤賓，蕤賓生大呂，大呂生夷則，夷則生夾鐘，夾鐘生無射，無射生仲呂。

三分所生益之一分以上生。

三分所生去其一分以下生。

黃鐘，大呂，太蔟，夾鐘，姑洗，仲呂，蕤賓，爲上。

林鐘，夷則，南呂，無射，應鐘，爲下。

The Huang-chung 黃鐘 pitch generates the Lin-chung 林鐘 pitch.
The Lin-chung pitch generates the T'ai-ts'ou 太蔟 pitch.
The T'ai-ts'ou pitch generates the Nan-lü 南呂 pitch.
The Nan-lü pitch generates the Ku-hsien 姑洗 pitch.
The Ku-hsien pitch generates the Ying-chung 應鐘 pitch.
The Ying-chung pitch generates the Jui-pin 蕤賓 pitch.
The Jui-pin pitch generates the Ta-lü 大呂 pitch.
The Ta-lü pitch generates the I-tsê 夷則 pitch.
The I-tsê pitch generates the Chia-chung 夾鐘 pitch.

The Chia-chung pitch generates the Wu-yi 無射 pitch.
The Wu-yi pitch generates the Chung-lü 仲呂 pitch.
The addition of 1/3 of its value gives rise to the up generation.
The subtraction of 1/3 of its value gives rise to the down generation.
Huang-chung, Ta-lü, T'ai-ts'ou, Chia-chung, Ku-hsien, Chung-lü, and Jui-pin are obtained by up-generation.
Lin-chung, I-tsê, Nan-lü, Wu-yi, and Ying-chung are obtained by down-generation.

The passage apparently describes a procedure for the generation of a set of twelve notes. The names of these notes are expressed in terms of the traditional Chou terminology. This suggests that the mechanism probably evolved from Chou.

Just like that evolved in the state Tseng, the Chou mechanism is also based on the up-and-down generation principle. In Figure 8, the Chou mechanism is illustrated generating twelve notes in the chromatic configuration. For convenience, the number 81 is adopted

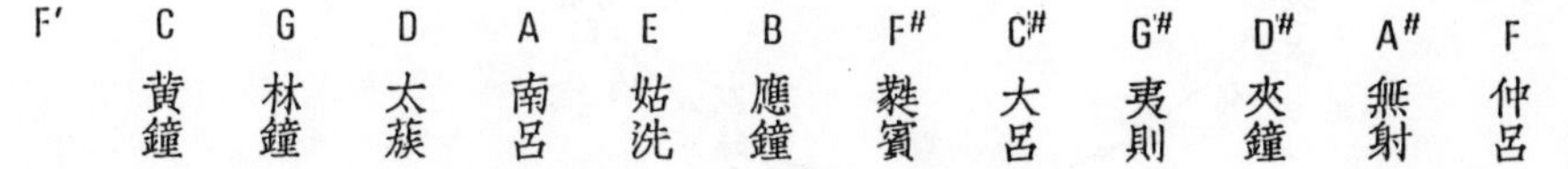

Figure 8. The generation of a chromatic scale by the up-and-down principle described in Chapter 6 of <u>Lü-Shih Ch'un-Ch'iu</u> 《呂氏春秋》 (<u>Master Lü's Spring-Autumn Annals</u>) of -239.

to represent the pitch 'Huang-chung' which is arbitrarily identified with the note C in the chromatic sequence. Again, it is understood that these notes are not tuned in accordance with the current convention of A - 440 Hz and their intervals are not tempered in equal temperament. Since, in this section, we are no longer dealing with the frequencies of notes directly, the multiplicative factors in the up-and-down principle are inverted back to their original forms (see Figure 4). The frequencies of notes are, of course, inversely proportional to the numbers generated by the principle and can be expressed in terms within a common factor ν_0. Thus, for example, the frequency of 'Huang-chung' can be written as $\nu_0/81$ and 'Wu-yi' as $729\ \nu_0/32768$.

The twelve notes generated by this procedure are found in the pattern of the well-known 'circle of fifths' as shown in Figure 9. It is

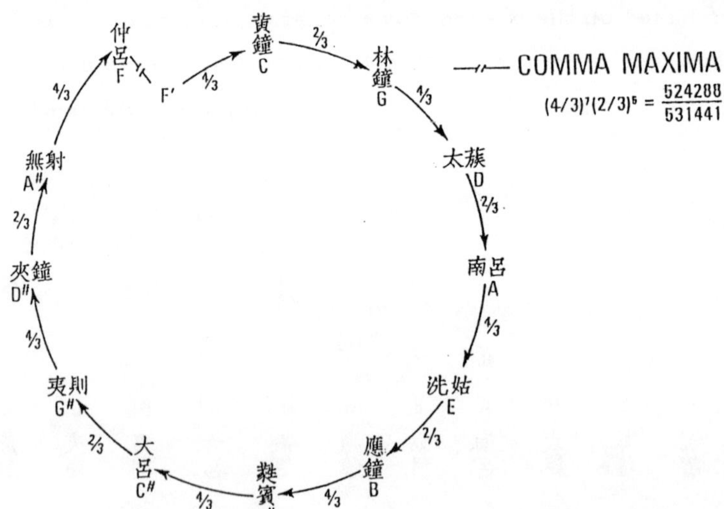

Figure 9. Circle of fifths generated by the method described in Chapter 6 of Lü-Shih Ch'un-Ch'iu 《呂氏春秋》 (Master Lü's Spring-Autumn Annals) of -239.

stated explicitly in the procedure that 'Huang-chung' 黃鍾 is obtained by up-generation. The note from which the note 'Huang-chung' is obtained is not the last note 'Chung-lü' 仲呂 (corresponding to the note F in the chromatic sequence), but a note whose name is not given (denoted by F′ in Figures 8, 9 and 10). The interval between these two notes is given by seven up-generations and five down-generations. We have

$$\left(\frac{4}{3}\right)^7 \left(\frac{2}{3}\right)^5 = \frac{2^{19}}{3^{12}} = \frac{524288}{531441}$$

which is the comma maxima. The number 262144/531551, one half of the comma maxima. has been incorrectly interpreted as the 'Chinese octave.'

The Chou procedure for the generation of a chromatic configuration has also been interpreted as the 'spiral of fifth.'[14] But the author has pointed out in an earlier paper[15] that there is no indication in the

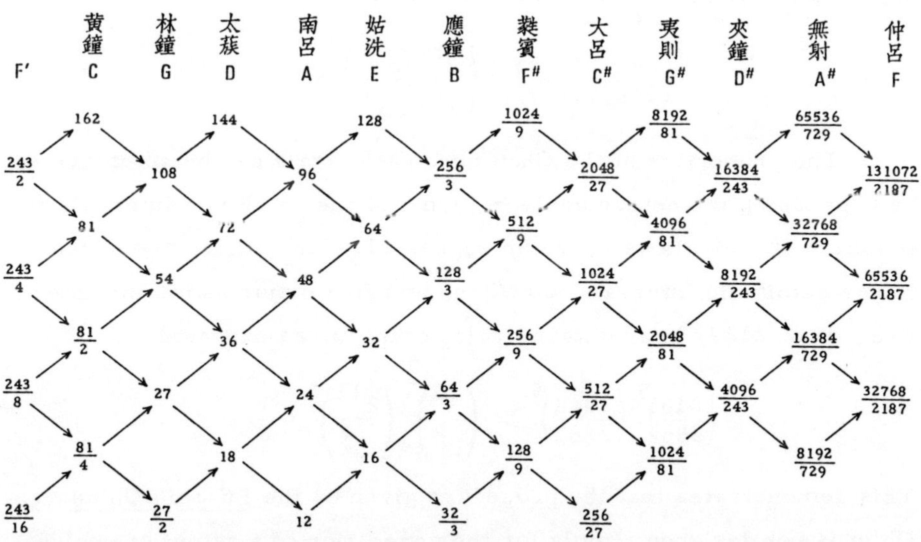

Figure 10. A net of linked chromatic configurations extending over three octaves.

passage from the Lü-Shih Ch'un-Ch'iu that either the octave or any of the notes outside of the octave are to be generated by the extension of the 'circle of fifths.'[16] In fact, the passage is very explicit in its choice of twelve notes lying within the compass of one octave.[17] To extend the chromatic configuration to the next octave, all one needs to do is to make use of the ratio 1/2 for an octave which is contained in the up-and-down principle (see Figure 4). In Figure 10, a net of such linked chromatic configurations extending over three octaves is given. These numbers reveal that the Chou chromatic scale is made up of two types of semitones, for example

$$\frac{\text{大呂}\,C^{\#}}{\text{黃鐘}\,C} = \frac{2048/27}{81} = \frac{2048}{2187} \quad \text{(major)}$$

$$\frac{\text{蕤賓}\,F^{\#}}{\text{仲呂}\,F} = \frac{512/9}{131072/2187} = \frac{243}{256} \quad \text{(minor)}$$

The combination of the minor semitone 243/256 and the major semitone 2048/2187 gives the major whole tone

$$\left(\frac{243}{256}\right)\left(\frac{2048}{2187}\right) = \left(\frac{3^5}{2^8}\right)\left(\frac{2^{11}}{3^7}\right) = \frac{8}{9}$$

The properties of the Chou chromatic scale can be summarized by expressing the octave cycle in terms of the semitonic intervals in the form of a circle as shown in Figure 11. Since there are seven minor semitonic intervals, 243/256, and five major semitonic intervals, 2048/2187, in an octave cycle, one has, as expected:

$$\left(\frac{243}{256}\right)^7\left(\frac{2048}{2187}\right)^5 = \left(\frac{3^5}{2^8}\right)^7\left(\frac{2^{11}}{3^7}\right)^5 = \frac{1}{2}$$

This demonstrates that the procedure given by the Lü-Shih Ch'un Ch'iu is not designed simply for the generation of a gamut of twelve pitches as previously assumed[12] but is a general mechanism for the generation of a chromatic scale.[15]

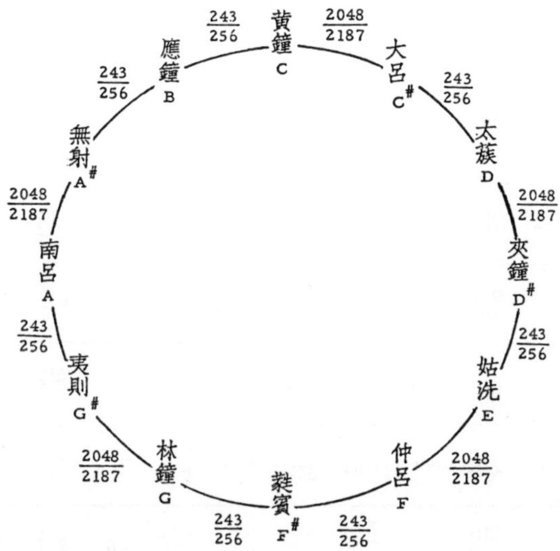

Figure 11. A circular representation of an octave cycle and its division into semitones for the Chou chromatic scale.

Literary evidence indicates that the early Chinese musicians were quite familiar with the 'hsüan-kung' 旋宮 principle, according to which each note of a scale can in turn act as the fundamental. We have, for example, from the Huai-Nan Tzu 《淮南子》(The Book of the Prince of Huai-Nan):

一律而生五音，十二律而爲六十音。

One pitch generates a set of five notes.
Twelve pitches form a set of sixty notes.

and from the Li Chi 《禮記》(Book of Chou Proprieties):

五聲，六律，十二管，旋相爲宮也。

[Each element of] a five-tone scale, a six-pitch scale and a scale of twelve pitch-pipes acts, in turn, as the fundamental.

Table VII. Comparison of the Intervallic Structure of the Twelve Keys of the Chou 周 Chromatic Scale and Natural (Just) Intonation.

Intervals \ Keys	黃鐘 C	大呂 C#	太蔟 D	夾鐘 D#	姑洗 E	仲呂 F	蕤賓 F#	林鐘 G	夷則 G#	南呂 A	無射 A#	應鐘 B	Natural (Just) Intonation
Fundamental	1	1	1	1	1	1	1	1	1	1	1	1	1
Minor second	$\frac{2048}{2187}$	$\frac{243}{256}$	$\frac{2048}{2187}$	$\frac{243}{256}$	$\frac{2048}{2187}$	$\frac{243}{256}$	$\frac{243}{256}$	$\frac{2048}{2187}$	$\frac{243}{256}$	$\frac{2048}{2187}$	$\frac{243}{256}$	$\frac{243}{256}$	$\frac{15}{16}$
Major second	$\frac{8}{9}$	$\frac{8}{9}$	$\frac{8}{9}$	$\frac{8}{9}$	$\frac{8}{9}$	$\frac{59049}{65536}$	$\frac{8}{9}$	$\frac{8}{9}$	$\frac{8}{9}$	$\frac{8}{9}$	$\frac{59049}{65536}$	$\frac{8}{9}$	$\frac{8}{9}$
Minor third	$\frac{16384}{19683}$	$\frac{27}{32}$	$\frac{16384}{19683}$	$\frac{27}{32}$	$\frac{27}{32}$	$\frac{27}{32}$	$\frac{27}{32}$	$\frac{16384}{19683}$	$\frac{27}{32}$	$\frac{27}{32}$	$\frac{27}{32}$	$\frac{27}{32}$	$\frac{5}{6}$
Major third	$\frac{64}{81}$	$\frac{64}{81}$	$\frac{64}{81}$	$\frac{6561}{8192}$	$\frac{64}{81}$	$\frac{6561}{8192}$	$\frac{64}{81}$	$\frac{64}{81}$	$\frac{6561}{8192}$	$\frac{64}{81}$	$\frac{6561}{8192}$	$\frac{64}{81}$	$\frac{4}{5}$
Perfect fourth	$\frac{131072}{177147}$	$\frac{3}{4}$	$\frac{3}{4}$	$\frac{3}{4}$	$\frac{3}{4}$	$\frac{3}{4}$	$\frac{3}{4}$	$\frac{3}{4}$	$\frac{3}{4}$	$\frac{3}{4}$	$\frac{3}{4}$	$\frac{3}{4}$	$\frac{3}{4}$
Augmented fourth	$\frac{512}{729}$	$\frac{729}{1024}$	$\frac{512}{729}$	$\frac{729}{1024}$	$\frac{512}{729}$	$\frac{729}{1024}$	$\frac{729}{1024}$	$\frac{512}{729}$	$\frac{729}{1024}$	$\frac{512}{729}$	$\frac{729}{1024}$	$\frac{512}{729}$	$\frac{32}{45}$
Perfect fifth	$\frac{2}{3}$	$\frac{2}{3}$	$\frac{2}{3}$	$\frac{2}{3}$	$\frac{2}{3}$	$\frac{177147}{262144}$	$\frac{2}{3}$	$\frac{2}{3}$	$\frac{2}{3}$	$\frac{2}{3}$	$\frac{2}{3}$	$\frac{2}{3}$	$\frac{2}{3}$
Minor sixth	$\frac{4096}{6561}$	$\frac{81}{128}$	$\frac{4096}{6561}$	$\frac{81}{128}$	$\frac{81}{128}$	$\frac{81}{128}$	$\frac{81}{128}$	$\frac{4096}{6561}$	$\frac{81}{128}$	$\frac{4096}{6561}$	$\frac{81}{128}$	$\frac{81}{128}$	$\frac{5}{8}$
Major sixth	$\frac{16}{27}$	$\frac{16}{27}$	$\frac{16}{27}$	$\frac{19683}{32768}$	$\frac{16}{27}$	$\frac{19683}{32768}$	$\frac{16}{27}$	$\frac{16}{27}$	$\frac{16}{27}$	$\frac{16}{27}$	$\frac{19683}{32768}$	$\frac{16}{27}$	$\frac{3}{5}$
Minor seventh	$\frac{32768}{59049}$	$\frac{9}{16}$	$\frac{9}{16}$	$\frac{9}{16}$	$\frac{9}{16}$	$\frac{9}{16}$	$\frac{9}{16}$	$\frac{32768}{59049}$	$\frac{9}{16}$	$\frac{9}{16}$	$\frac{9}{16}$	$\frac{9}{16}$	$\frac{5}{9}$
Major seventh	$\frac{128}{243}$	$\frac{128}{243}$	$\frac{128}{243}$	$\frac{2187}{4096}$	$\frac{128}{243}$	$\frac{2187}{4096}$	$\frac{128}{243}$	$\frac{128}{243}$	$\frac{2187}{4096}$	$\frac{128}{243}$	$\frac{2187}{4096}$	$\frac{128}{243}$	$\frac{8}{15}$
Octave	$\frac{1}{2}$	$\frac{1}{2}$	$\frac{1}{2}$	$\frac{1}{2}$	$\frac{1}{2}$	$\frac{1}{2}$	$\frac{1}{2}$	$\frac{1}{2}$	$\frac{1}{2}$	$\frac{1}{2}$	$\frac{1}{2}$	$\frac{1}{2}$	$\frac{1}{2}$

This implies that the Chou chromatic scale can be expressed in twelve possible keys. In Table VII the intervallic structure of the Chou chromatic scale is given in the twelve keys. Also included in Table VII is the intervallic structure of natural (just) intonation.

Apparently the Chou procedure for generating the chromatic scale is more advanced than the Tseng procedure. A comparison of the intervallic structure of the Tseng chromatic scale with that of the Chou chromatic scale is given in Table VIII. For comparison, the intervallic structures of natural and equal-temperament intonation are also included. In this comparison, one must take into consideration that the entries for the Tseng scale are derived from the frequency measurements of the bells. They are not theoretical values. In general, the intervals in the Tseng chromatic scale as represented by the bells are larger than those of the Chou chromatic scale. This probably reflects, to a large extent, the practical difficulty in reproducing the scale in bells. It should be noted that if the intervals represented by 'chio' 角 and by 'tseng' 曾, in the Tseng procedure, were indeed the ratios 81/64 and $(81/64)^2$, respectively, then the calculated intervallic structure by the Tseng procedure would be identical to that obtained by the Chou procedure.

The claim that early Chinese musicians knew about the 'hsüan-kung' principle is supported by the inscriptions found on the bronze set-bells of the -5th century. Among the 13 'niu chung' set-bells, nine bells have inscriptions which provide, in addition to the names of their notes, names of the keys when their notes act as the fundamental. For example, the note 'kung-chio' 宮角 of bell Top 3-6 when acting as the fundamental gives the 'T'ai-ts'ou' 太蔟 [大族] key. The most interesting case is the note 'shang' 商 of bell Top 2-6 which, when acting as the fundamental, gives the 'Huang-chung' 黃鍾 key. This implies that the frequency of the Huang-chung

Table VIII. A Comparison of the Intervallic Structure of the Tseng 曾 and Chou 周 Chromatic Scales.

Intervals	Natural Intonation (cents)	Tseng 曾 Intonation (cents)	Chou 周 Intonation[a] (cents)	Tempered Intonation (cents)
Fundamental	0	0	0	0
Minor second	111.73	147.61	113.69	100
Major second	203.91	243.01	203.91	200
Minor third	315.64	344.87	317.60	300
Major third	386.31	487.56	407.82	400
Perfect fourth	498.05	573.62	521.51	500
Augmented fourth	590.22	656.76	611.73	600
Perfect fifth	701.96	777.53	701.96	700
Minor sixth	813.69	884.35	815.64	800
Major sixth	884.36	965.74	905.87	900
Minor seventh	1017.60	1002.80	1109.55	1000
Major seventh	1088.27	1197.25	1109.78	1100
Octave	1200	1205.27	1200	1200

[a] In the 'Huang-chung' key.

fundamental at the time of the -5 century was 410.1 vibs/sec, as given by the measured frequency for the note 'shang' 商 of bell Top 2-6. This permits a direct comparison of the frequencies of the Tseng and the Chou chromatic scales in the 'Huang-chung—kung' 黃鍾—宮 mode-key.

Such a comparison is given in Table IX. Included in Table IX is the natural intonation. The notes of the Tseng chromatic scale are consistently of higher frequencies than both the Chou and natural intonations. The frequency for the major seventh is, however, much too large and appears to be out of the general characteristics of the

Table IX. A Comparison of the Frequencies of the Tseng 曾 and Chou 周 Chromatic Scales in the 'Huang-chung—kung' 黃鍾—宮 Mode-Key with Natural (Just) Intonation.[a]

Notes	Tseng Intonation (vibs/sec measured)	Chou Intonation (vibs/sec calculated)	Natural (Just) Intonation (vibs/sec calculated)
Fundamental	410.1	410.10	410.10
Minor second	446.6	437.93	437.44
Major second	471.9	461.36	461.36
Minor third	500.5	492.68	492.12
Major third	543.5	519.03	512.63
Perfect fourth	571.2	554.26	546.80
Augmented fourth	599.3	583.91	576.70
Perfect fifth	642.6	615.15	615.15
Minor sixth	683.5	656.90	656.16
Major sixth	716.4	692.04	683.50
Minor seventh	731.9	739.01	738.18
Major seventh	818.9	778.55	768.93
Octave	822.7	820.20	820.20

[a] Normalized to the actual frequency, 410.1 vibs/sec, of the 'Huang-chung' fundamental in accordance with the inscription found on bell Top 2-6 unearthed in 1978 from the tomb of Marquis Yi of the Tseng state found in Sui county in Hupei province. The tomb was sealed around -433.

Tseng intonation. There are a number of possible factors behind this anomaly that may have nothing to do with the theoretical aspect of the Tseng procedure. It is worthwhile to re-examine bell Top 2-4 for any pecularities. In view of the fact that the Tseng set-bells were cast over 2400 years ago, the comparison given in Table IX is amazingly good.

IV. Concluding Remarks

The discovery of the Tseng musical instruments, especially the bronze set-bells, in Marquis Yi's tomb have greatly enhanced our knowledge of early Chinese work on musical instruments, tonal systems, and musicological theory. It has provided not only a convincing proof for the early development of the chromatic scale in China but also evidence for discrediting a number of incorrect assertions on ancient music in Chinese civilization.

An excellent account of early discussions of Chinese music from the publication of Amiot's work in 1780[18] to the work of Chavannes and up to the early 20th century[19] is given by Needham and Robinson.[20] I will not repeat it here. The turn of events from the extreme pro position of Amiot to the extreme con position of Chavannes led to the characterization that Chinese music was monophonic and pentatonic and the Chinese musical scale was impure containing no perfect fourth, semitones or the true concept of an octave.

This characterization of the Chinese musical scale must now be dismissed. Not only did the perfect fourth, semitones and the perfect octave appear in Chinese musical scales, the concept of perfect fourth and octave were incorporated in the up-and-down principle in their mathematical procedure for the generation of scales. The appearance of semitones in Chinese music can be traced back to a date much earlier than that afforded by the Tseng set-bells. The earliest extent evidence of semitones is found in a pottery globular flute ('hsüan' 塤) unearthed from a Yin 殷 tomb of the -11th century at Liuliko 琉璃閣 in Hui 輝 county, Honan 河南 province. The globular flute is the only undamaged one of a set of three unearthed in the tomb. It has five asymmetrically placed finger holes and is capable of producing a set of eleven tones. Measurements carried out by the Ethnic Music Research Institute of the Central Music Conservatory in Peking reveal that nine of the eleven tones are identifiable with the semitones of a chromatic scale of which seven are consecutive semitones within an octave.[21] This seems to suggest that a twelve-note scale was probably in the making by the Shang 商 period.

There is no doubt that the early Chinese musicians made extensive use of pentatonic intonation for melodic articulation in music. This does not imply that they made no use of other intonations. In fact, the early Chinese musicians did make use of hexatonic, heptatonic and chromatic intonations to harmonize and modulate pentatonic intonation. We have, for example, from the Tso Chuan 《左傳》 (Master Tsoch'iu's Commentary on the Spring-Autumn Annals) the statement:

七音，六律，以奉五聲。

> The seven-note scale and the six-pitch scale subordinate to the five-tone scale.

and from the Lü-Shih Ch'un-Ch'iu the statement:

鑄十二鐘以和五音。

> The set of twelve bells is cast in order to harmonize five-note scale.

It is worthwhile to note that before the word 'sheng' 聲 and the word 'yin' 音 became synonyms by the late -3rd century, they had a subtle difference. An old definition of these two words is found in Chapter 24 of the Shih Chi 《史記》 (Historical Records). We have

單出曰聲，雜比曰音。

> A singly emerged tone is called 'sheng.'
> An assorted comparative sound is called 'yin.'

Thus the term 'wu sheng' 五聲, which means literally five singly emerged tones, denotes specifically the five tones that make up the pentatonic intonations. Intonations with auxiliary tones added for harmonic purposes are referred to as 'yin.' For example, the heptatonic intonation containing two additional notes as auxiliary tones in bridging the two one-and-half step gaps in the pentatonic intonation were called 'ch'i yin' 七音. It is in the Lü-Shih Ch'un-Ch'iu that the use of 'sheng' and 'yin' interchangeably is first observed.

The early Chou musicians were particularly partial to the pentatonic intonation since it is composed of 'sheng' 聲. In their music, 'sheng' plays the major role while the 'yin' plays subordinate roles. A description of the role of 'sheng' and its relationship to 'lü' 律 is found in the Kuo Yü 《國語》 (State Discourses):

聲以和樂，律以平聲。

'Sheng' is used to harmonize music.
'Lü' is used to regulate 'sheng.'

The description of the relationship between 'lü' and 'sheng' is of particular interest since each pitch is capable of generating a set of 'sheng.' This seems to suggest the 'hsüan-kung' 旋宮 principle.

Another question of interest concerns the method by which the early Chinese musicians determined the ratios of harmonic tones that formed the bases for the generation of their musical scales. Before frequency of vibration and its relation to sound was known, early musicians worked with the concept of pitch in terms of the length of the resonating agent that produced the pitch. The intervallic ratios can be determined from the ratios of the lengths of a stretched string adjusted to reproduce such harmonic tones. There has been much debate among historians of science on the question of whether the early Chinese made use of such a method. One of the objections to the Chinese use of the method of a stretch string is based on the absence of epigraphic evidence or artifacts of any early Chinese string instruments even though string instruments such as 'se' 瑟 (zither) and 'ch'in' 琴 (half-tube zither) are frequently mentioned in the early Chou classics such as the Shih Ching 《詩經》(Book of Odes). With the discovery of string instruments in the Marquis Yi's tomb, we now have string instruments dating back to the -5th century. Two string instruments, a five-string 'se' and a ten-string 'ch'in,' were found in the tomb. These are rather advanced string instruments, suggesting that the development of string instruments in Chinese civilization began probably even earlier.

There are also a number of historians who consider that the Chinese received their ratios from other civilizations. Their arguments are based on a calculation for the construction of a set of twelve pitch-pipes found in the Shih Chi of -90. In this calculation, the ratios of pipe lengths were taken to be the corresponding pitch ratios. This is, of course, incorrect since there must be end-corrections to the pipe lengths. Such a use of pitch ratios has been considered by some historians of science to be an indication of a lack of understanding of pitch ratios. Hence, it supports their view that the Chinese received their ratios from other civilizations.

It is certainly true that if the pitch-pipes were constructed based on the harmonic pitch ratios, end-corrections in pipe length would then be needed. We now know that the actual nodal point lies a distance out from the open ends of the pipe. Thus, the actual length of the resonating air column in the pipe is somewhat longer than the pipe length itself.

In examining this calculation, one must put the problem in its proper context. The calculation given in the Shih Chi was not an attempt to demonstrate the relationship between harmonic pitch ratios and the length of a resonance agent. It was an attempt to generate a set of twelve standard pitches. For this purpose it automatically precluded the use of a stretched string despite the fact that the ratios in string length are exactly equal to the corresponding ratios in pitch. The reason is simply because the fundamental pitch of a stretched string depends also on the tension and weight of the string. To maintain the pitch as a standard, a constant tension in the string must be maintained. The maintenance of the string tension would have given rise, in ancient times, to practical difficulties. The fundamental frequency of a stretched string is proportional to the square root of its tension.

Although the ratios of pipe lengths are not given exactly by the pitch ratios, the use of pitch-pipes for the standardization of pitch was a practical choice of scientific merit at that stage in its development, since the end corrections are relatively small for pipes with small diameters and once the pipe length is fixed, the pitch would not alter appreciably under normal conditions. This is because, as we now know, the other factors such as temperature, humidity and density of the air column which may shift the fundamental frequency would not vary under normal conditions to the extent of producing large alterations in the pitch. The higher partials of the harmonics that can be produced due to wind-pressure increase through overblowing should not pose much difficulty for trained musicians.

The early interest in the standardization of pitch probably came from the desire to play different musical instruments together for a greater variety in timbre and tone color to enrich their musical sound. According to the Shih Ching, the practice of playing several instruments together began very early in China. In order to do so, one must deal with the problem of tuning when musical instruments with unadjustable pitch such as the globular flute, set-bells, and reed mouth-organ ('sheng' 笙) are played together. In addition, the ability to play differently keyed musical instruments together depends also on the compatibility of their tonality with respect to modulation.[22] This undoubtedly helped the early Chinese musicians in the development of musicological theories such as the 'hsüan kung' 旋宮 principle.

In their search for a durable and authoritative instrument for the standardization of pitch, the Chinese musicians also made use of bronze set-bells, as the art of campanology had reached a stage of maturity. The Tseng set-bells unearthed from the Marquis Yi's tomb constitute such an example. In the history of campanology in China, the claim that bells were invented by Huang Ti 黃帝, a pre-historic ruler of the -3rd millennium, has met with skepticism among historians. A common reaction to this claim is that it is simply not possible, since bronze casting had not yet been developed at that time.[23] But if one examines the early records, one would find that no claim is made that the early bells were cast in bronze. Archeological finds reveal that bell-making did begin in China early in the Neolithic period but that the early bells were made of clay.[24] During the period of the Lungshan 龍山 culture (c. -2800 to c. -2100), contemporary to the traditionally assigned Huang-Ti period, a new type of pottery bell began to emerge. In Figure 12, a clapperless pottery bell ('t'ao chung' 陶鐘) of the Lungshan culture is shown. Included for comparison in Figure 12 is a bronze bell of the Yin 殷 period (c. -1384 to c. -1111) of the Shang 商 dynasty. The comparison reveals that the bronze bells inherited their design from the pottery bells of the -3rd millennium. Thus, the development of campanology in Chinese civilization went through a pottery stage before bronze bells were cast. The earliest extant bronze bells come from the early Shang period of the -17th century. It took over a millennium for this evolutionary process to culminate in the bronze set-bells of Tseng.

Figure 12. A comparison of a clapperless pottery bell of the -3rd millennium with a clapperless bronze bell of the Yin 殷 period (c-1384 to c.-1111) of the Shang 商 dynasty. Left, unearthed from a Lung-shan 龍山 cultural site located at K'asheng-chuang 客省莊 in Ch'ang-an 長安, Shensi 陝西 provine. Right, unearthed from a Yin tomb at Taszuk'ung 大司空 near Anyany 安陽, Honan 河南 province.

Acknowledgments. I am grateful to Lam Ping-Liang 林炳良 for his encouragement of this work and to Xi Ze-Zong 席澤宗 for making available to me Wang Hsiang's 王湘 personal copy of the special issue of the 'Yin-Yüeh Yen-Chiu' 《音樂研究》(Music Research) devoted to works on the Tseng 曾 musical instruments unearthed from Marquis Yi's 乙 tomb.

References

1. See the report of the archeological team of the first tomb at Leikutun 擂鼓墩 in Sui 隨 county, 'Hupei Sui-Hsien Tseng-Hou-Yi-Mu Fa-Chüeh Chien-Pao' 《湖北隨縣曾侯乙墓發掘簡報》 ('A Brief Report on the Excavation of the Tomb of Marquis Yi of Tseng') in Wen Wu 《文物》 (Cultural Relics) 7, 1-16 (1979).

2. Huang Hsiang-P'eng 黃翔鵬, 'Hsien-Ch'in Yin-Yüeh Wen-Hua ti Kuang Hui Ch'uang-Tsao' 〈先秦音樂文化的光輝創造〉 ('Great Achievements in the Pre-Ch'in Musical Culture') in Wen Wu 《文物》 (Cultural Relics) 7, 32-39 (1979).

3. 'Sui Hsien Tseng-Hou-Yi-Mu Chung-Ch'ing Ming-Wen Shih-Wen' 〈隨縣曾侯乙墓鐘磬銘文釋文〉 (Interpretation of the Inscriptions on the Bells and Chimes in the Tomb of Marquis Yi of Tseng in Sui County') by the Museum of Hupei Province in Yin-Yüeh Yen-Chiu 《音樂研究》 (Music Research) 1, 3-16 (1981).

4. Ch'iu Hsi-Kuei 裘錫圭 and Li Chia-Hao 李家浩, 'Tseng-Hou-Yi-Mu Chung-Ch'ing Ming-Wen Shih-Wen Shuo-Ming' 〈曾侯乙墓鐘磬銘文釋文說明〉 ('Exposition of the Interpretation of the Inscriptions on the Bells and Chimes of the Tomb of Marquis Yi of Tseng') Yin-Yüeh Yen-Chiu 《音樂研究》 (Music Research) 1, 17-21 (1981).

5. Huang Hsiang-P'eng 黃翔鵬, 'Tseng-Hou-Yi Chung-Ch'ing Ming-Wen Yüeh-Hsüeh T'i-Hsi Ch'u-T'an' 〈曾侯乙鍾磬銘文樂學体系初探〉 ('A Preliminary Study of Musicological Terminology in the Inscriptions on the Bells and Chimes of Marquis Yi of Tseng') Yin-Yüeh Yen-Chiu 《音樂研究》 (Music Research) 1, 22-53 (1981).

6. Li Ch'un-I 李純一, 'Tseng-Hou-Yi Pien-Chung Ming-Wen K'ao-So' 〈曾侯乙編鍾銘文考索〉 ('Investigation of the Inscriptions on the Set-Bells of Marquis Yi of Tseng') Yin-Yüeh Yen-Chiu 《音樂研究》 (Music Research) 1, 54-67 (1981), and 'Tseng-Hou-Yi-Mu Pien-Ch'ing Ming-Wen Ch'u-Yen' 〈曾侯乙墓編磬銘文初研〉 ('A Preliminary Study of the Inscriptions on the Stone Set-Chimes of the Tomb of Marquis Yi of Tseng') Yin-Yüeh Yi-Shu 《音樂藝術》 (Art of Music) 1(12), 8-24 (1983).

7. Wang Hsiang 王湘, 'Tseng-Hou-Yi-Mu Pien-Chung Yin-Lü ti T'an-T'ao' 〈曾侯乙墓編鐘音律的探討〉 ('An Exploration of the Temperament of the Set-Bells of the Tomb of Marquis Yi of Tseng') Yin-Yüeh Yen-Chiu 《音樂研究》 (Music Research) 1, 68-78 (1981).

8. P'an Chien-Ming 潘建明, 'Tseng-Hou-Yi Pien-Chung Yin-Lü Yen-Chiu' 〈曾侯乙編鐘音律研究〉 ('Research on the Temperament of Marquis Yi's Set-Bells') in the Thirty-Year-Anniversary Special Edition of Shanghai Po-Wu-Kuan Chi-K'an 《上海博物館集刊》 (A Special Issue of Collected Essays of the Shanghai Museum) (Old Book Publ. Co., Shanghai, 1982), pp. 93-115.

9. T'an Wei-Szu 譚維四 and Feng Kuang-Sheng 馮光生, 'Kuan-Yu Tseng-Hou-Yi-Mu Pien-Chung Niu-Chung Yin-Yüeh Hsing-Neng ti Chien-Chien' 〈關于曾侯乙墓編鐘鈕鐘音樂性能的淺見〉 ('On the Observations of the Musical Nature of the Set-Bells with Knobs from the Tomb of Marquis Yi of Tseng') Yin-Yüeh Yen-Chiu 《音樂研究》 (Music Research) 1, 79-87 (1981).

10. 'Tseng-Hou-Yi Pien-Chung P'in-Lü Shih-Ts'e' 〈曾侯乙編鐘頻率實測〉 ('Laboratory Measurement of the Frequencies of

Marquis Yi's Set-Bells of the Tseng') by the Bronze Research Division of the Shanghai Museum in the Thirty-Year-Anniversary Special Edition of Shanghai Po-Wu-Kuan Chi-K'an 《上海博物館集刊》(A Special Issue of Collected Essays of the Shanghai Museum (Old Book Publ. Co., Shanghai, 1982), pp. 89-92.

11. See for example, Chapter 21A of the Ch'ien Han Shu 《前漢書》 (History of the Former Han Dynasty). We have the statement:

律有十二，陽六爲律，陰六爲呂。

There are twelve pitches: Six 'yang' 陽 pitches compose the 'lü' 律 system and six 'yin' 陰 pitches compose the 'lü' 呂 system.

12. Joseph Needham and Kenneth Robinson, 'Sound (Acoustics)' in Science and Civilisation in China (The University Press, Cambridge, 1962), vol. 4, sec. 26, pp. 126-228.

13. A. Forke, Geschichte d. alten chinesischen Philosophie (De Gruyter, Hamburg, 1927), p. 74.

14. Joseph Needham and Kenneth Robinson, op. cit. (ref. 12), p. 212.

15. Chen Cheng-Yih 程貞一, 'Early Chinese Work on Harmonic Progression in Tonal System' in the Proceedings of the 3rd International Conference on the History of Chinese Science, August 20-24, 1984 (Science Publ. Co., Peking, to be published).

16. The first attempt at extending the 'circle of fifths' into a 'spiral of fifths' was made by Ching Fang 京房 (b.-77, d.-37). He extended the up-and-down principle from 12 to 60 and obtained 60 microtonic intervals within the compass of an octave. The work, however, was not an attempt to determine notes outside

of an octave or to determine the octave. See Hou Han Shu 《後漢書》 (History of the Later Han Dynasty).

17. The Chou procedure allows two possible choices of a chromatic configuration with all twelve notes lying within the compass of an octave. An alternative choice of the twelve notes is to replace the last generated note, 'chung-lü' 仲呂, by the unnamed note (denoted by F' in Figures 8, 9 and 10) from which the note, 'huang-chung' 黃鍾 is generated. In view of the cyclic nature of the scale, the two choices are in fact equivalent with respect to the overall structure, but the structure of the individual mode-keys will be interchanged. The choice made by the Chou musicians has the distinction of having a slightly sharper 'perfect' fourth, 131072/177147, in the set of 'huang-chung' mode-keys but, in all of the other eleven possible mode-keys, the perfect fourth remains at 3/4 (see Table VII). Note that (131072/177147) = (3/4) - (7153/708588).

18. Joseph J.-M. Amiot, 'Mémoire sur la Musique des Chinois tant anciens que modernes,' Mémoires concernant l'Histoire, les Sciences, les Arts, les Maeurs et les Usages, des Chinois, par les Missionnaires de Pékin, Paris, 1776-1814. 6, 1 (1780) (written in 1776).

19. E. Chavannes, Les Memoires Historiques de Se-Ma Ts'ien (Leroux, Paris, 1895-1905), vol. 3.

20. Joseph Needham and Kenneth Robinson, op. cit. (ref. 12), p. 176.

21. Li Ch'un-I 李純一, Chung-Kuo Ku-Tai Yin-Yüeh Shih-Kao 《中國古代音樂史稿》 (A Sketch of the History of Ancient Chinese Music (Music Publ. Co., Peking, 1964).

22. An investigation of modulation among both pentatonic and heptatonic mode-keys with the Chou chromatic scale is given by Chen, see Ref. 15.

23. Based on the present state of our archeological knowledge, there is now little doubt that the emergence of bronze technology began in China no later than the -3rd millennium. See Yan Wen-Ming 嚴文明, 'Lun Chung-Kuo ti T'ung-Shih-Ping-Yung Shih-Tai'〈論中國的銅石併用時代〉('On the Chalcolithic Age of China') in Shih-Chien Yen-Chiu《史前研究》Prehistory 1, 36-44 (1984) and Ko Tsun 柯俊, 'The Development of Metal Technology in Ancient China' in this book, p. 225.

24. Hsia Nai 夏鼐, K'ao-Ku Hsüeh Lun-Wên-Chi《考古學論文集》(Collected Essays on Archeology) (Science Publ. Co., Peking, 1961), p. 28.

NEW SOURCE MATERIALS FOR METEOROLOGICAL INSTRUMENTS IN ANCIENT CHINA

Di Li 李迪

Ancient China, being an agriculturally oriented society, had a deep interest in weather matters from antiquity and kept records of weather over a much longer period of time than most other ancient civilizations. However, there were only a few attempts to provide an account of the history of meteorology in ancient China.[1] Much effort has recently been directed to a systematic search of ancient records on weather matters.[2] In this paper, I report some recently discovered records concerning two types of meteorological instruments, namely the wind gauge and the rain gauge.

I. The Wind Gauge

Determination of wind direction by means of a piece of silk or cloth hung on the top of a pole such as a weather vane began in antiquity. It is perhaps the oldest of all pointer-reading devices. The question that is of interest in the development of the wind gauge is when such a device was replaced by the 'xiang feng wu' 向風烏 (weather crow), the Chinese counterpart of the weather cock in the West. Not until the discovery of a weather crow on a wall painting in a tomb of the Eastern Han 漢 dynasty in 1972 was it possible to make any kind of a definite answer to this question.[3] The epigraphic evidence provided by the wall painting indicates that the weather crow was invented, at the latest, by the Han dynasty. It should be noted that traditional accounts found, for example, in the Gu Jin Zhu 《古今注》 (Commentary on Things Old and New) of the 4th century[4] and in the Shi Wu Ji Yuan 《事物紀原》 (Records of the Origins of

Affairs and Things) of c. 1085 place the invention at a much earlier date.

In a book written shortly after the Han dynasty in the 3rd century, San Fu Huang Tu 《三輔黃圖》(Description of the Three Districts in the Capital), it is stated that on an observatory of the Han dynasty, a copper weather crow was mounted for the measurement of wind direction.

The weather crow was such a simple instrument that it soon became a common tool found in cities and villages. Illustrations of weather crows are found in popular Chinese paintings such as the 'Qing Ming Shang He Tu' 清明上河圖 ('Riverside Scene at the Spring Festival') by the well-known painter Zhang Ze-Duan 張擇端 of the Song 宋 dynasty. The illustrations (four in total) indicate a fixed horizontal wooden cross was placed below the bird to act as a reference for determination of the direction of the wind indicated by the bird. The weather crows in this picture are placed on top of poles located on the two sides of bow-bridges.

The weather crow also found its way into divination as described in books such as the Yi-Si Zhan 《乙巳占》(Yi-Si Divination) of the early Tang 唐 dynasty and the Guan-Xiang Wan Zhen 《觀象玩占》(Prognostication by Weather in the Game of Divination).

A technical term for wind gauge was 'feng lun' 風輪 (literally 'wind wheel'). The term is found in a passage in the Mi Shu Jian Zhi 《祕書監志》(Records of the Secretary's Department [of the Yuan court]) dated 1276. We have:

> According to the former practice of the observatory, a 'feng lun' should be mounted at the top of the observatory so that the wind in its eight directions can be plotted as a function of time. The head of the secretary's department has paid several inspection

> visits to our observatory and asked that it be installed as in the past. Until now the request has not been accomplished. Director of the observatory, please appoint a third-class officer to be the person in charge so that it can be constructed and mounted in a few days. Persons should then be appointed to keep records of the weather and append it to the calendar as in the past. If there is any delay the cause must be determined and appropriate action taken.

The observatory mentioned in the passage was that in the captial Dadu 大都 of the Jin 金 dynasty and was called the 'Si Tien Tai' 司天臺 at that time.

In the seventeenth century, a Western weather cock was mounted on the observatory in Beijing but Chinese traditional weather crows did not disappear. The weather crow was incorporated into the tool manual, He Gong Qi Ju Tu Shuo《河工器具圖說》(River Tools Illustrated) of 1836 by Lin Qing 麟慶. On April 6, 1986, the author had the opportunity to examine an ancient weather crow still in use at Hunyuan 渾源, Shanxi 山西 (see Figure 1). The weather crow is made of iron and mounted at the top of a Buddhist temple.

Figure 1. A photograph of a traditional Chinese wind gauge 'xiang feng wu' 向風烏 made of iron mounted on a Buddhist temple at Hunyuan 渾源 county, Shanxi 山西.

II. Rain Gauges and Evaporators

It is now generally recognized that the invention of the rain gauge in China dates from a time probably a good deal earlier than the Song dynasty. In any event, the invention could not be later than the work of Qin Jiu-Shao 秦九韶, for in his book Shu Shu Jiu Zhang 《數書九章》(Mathematical Treatise in Nine Sections) of 1247, we find problems which are designed to teach one the use of the rain gauge and snow gauge. The Chinese technical name for the rain gauge is 'ce yu qi' 測雨器 ('rain-measuring instrument'). Additional information on Chinese work on rain gauges comes indirectly from the Korean historical records of the 15th and 18th centuries.[1,5]

In this section, I report on a new Chinese source for the rain gauge found in an unpublished manuscript entitled Ce Yuan Tu Jie《測圓圖解》(Illustrated Analysis of Circle Measurements) by Dai Yuan 戴源 of the 18th century, recently discovered in the Zhejiang 浙江 Library. The text entitled 'The Method for Measuring Annual Rainfall' contains three passages. We have, as the first passage:

> Construct a large copper or tin box with length, width and height all 3 'chi' 尺 (Chinese feet) (see Figure 2). A hole is made at a lower corner of the box in which a copper tube is inserted. The copper tube must be properly fitted with a wooden stop to prevent leakage. Another small box of length 3 'cun' 寸 (Chinese inches), width 3 'cun' and height 1 'cun' is then made (see Figure 2). Put the larger box in a courtyard so that it will be filled with rainwater when it rains. After the rain stops remove the wooden stopper and allow the water in the box to flow into the smaller box. Count the number of times the small box must be emptied to drain the large box. In the case that the large box is full and it is still raining, the small box should be used to drain the rainwater until the large box is either completely or half emptied. Then plug the copper tube with the wooden stopper. Make sure to mark down every time the small box is filled this way.

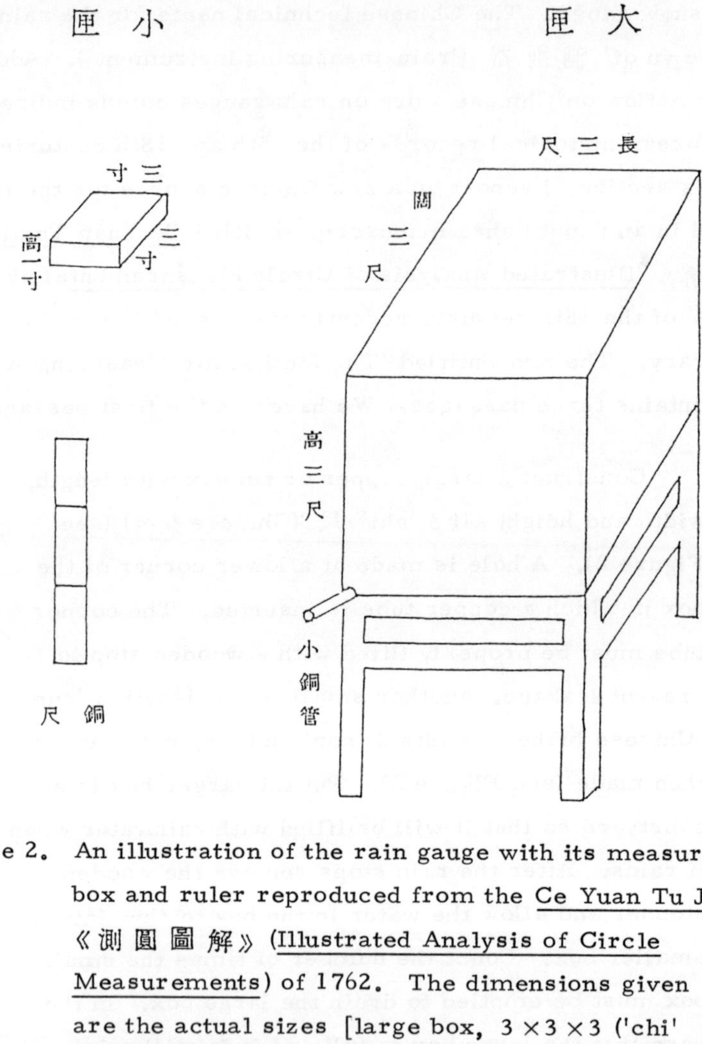

Figure 2. An illustration of the rain gauge with its measuring box and ruler reproduced from the <u>Ce Yuan Tu Jie</u> 《測圓圖解》(<u>Illustrated Analysis of Circle Measurements</u>) of 1762. The dimensions given are the actual sizes [large box, $3 \times 3 \times 3$ ('chi' 尺)^3, small box $3 \times 3 \times 1$ ('cun' 寸)^3, 1 'chi' = 32 cm, 1 'cun' = 3.2 mm].

> The method for computing the rainfall is as follows. If the amount of water is less than 10 small boxes, it is not necessary to record. We get 1 'fen' 分 (one tenth of a Chinese inch) of rainfall for 10 small boxes of water. One hundred boxes give 1 'cun' of rainfall while 1000 boxes gives 1 'chi' of rainfall. There is no need to use the small box to measure the water if the larger box is filled level to the top, since the rainfall is just three 'chi' as the volume of the larger box is 3000 times that of the smaller one.

This passage provides specific instructions for both the construction of a rain gauge and the determination of the amount of rainfall measured by the rain gauge.

The second passage from the manuscript deals with an evaporator. We have:

> Construct a box of the same size as the large rain gauge box but with no hole in the box. Place the box also in a courtyard and then fill it with water level to the top. When it is raining, the box must be covered with a board so that no rainwater will get into it. Make a copper ruler 3 'chi' long to match the height of the box. Divide the ruler evenly into 30 'cun' with each 'cun' divided again evenly into 10 'fen.' Measure the depth of water remaining in the box using the ruler every month or half a month so as to observe how much water has evaporated. Each observation should be recorded. If the water remaining in the box is half

> or less than half of its full amount, additional water should be added to raise the water again to the top. The amount added should be measured and recorded each time.

Finally, we have the third passage:

> The two large boxes (one for the rain gauge and the other for the evaporator) should be placed on a wooden stool of height approximately 2 'chi' and properly leveled. Every season the data obtained from the two instruments should be compared to evaluate the season's dryness or wetness. The results obtained over a year taken together should provide a reasonable overall picture of the dryness or wetness of the region.

The account of a rain gauge and evaporator given by these three passages is of particular interest since it is the first record currently known in China of the use of the two instruments together for evaluation of the actual availability of water. It should be noted that, according to the meteorological system at the time of Dai Yuan in the 18th century,[6] 1 'cun' was equal to 3.2 cm. Hence 1 'fen' equalled 3.2 mm. This implies that the volume of the small box of the rain gauge was approximately 295 cm^3. Based on the description given in the first passage, the amount of water in 10 such small boxes (i.e., 2950 cm^3) corresponds to a rainfall of 3.2 mm. Perhaps, therefore, rainfall of less than 10 small boxes was not recorded as the amount of rain was considered negligible.

The Dai Yuan's manuscript discovered at the Zhejiang Library is hand-written and unpublished. It contains mainly mathematical work but includes also works in physics, surveying, astronomy, meteorology and other fields. These works are not divided into chapters. There is no preface or postscript in the manuscript. At the end of the section on proportional dividers, it is stated that Dai Yuan of Qiantang 錢塘 witnessed it in December of the twenty-sixth year of the Qian Long 乾隆 reign. This implies that the work was written in December of 1762.[7] According to Li Yan 李儼 (1892-1963), a well-known historian of Chinese mathematics, Dai Yuan wrote a book known as the Dai-Shi Suan-Shu《戴氏算術》(Dai Yuan's Mathematical Manual), and the record also states that the book in the form of a manuscript with a preface written by Dai Yuan himself was preserved by Liao Shi-Yin 廖世蔭 and later by Gu Ming 顧明.[8] The Dai Shi Suan-Shu is probably just the Ce Yuan Tu Jie.

By the 18th century, Western influence on science in China was significant. It is natural to ask whether the rain gauge and evaporator described by Dai Yuan were transmitted from the West. Dai certainly knew Western science since he made reference to Western work in his manuscript. However, in the section dealing with the rain gauge and evaporator, he made no reference to Western work. In Europe, the rain gauge was first made in 1639 by Benedetto Castelli at Perugia. According to Liu Zhao-Ming 劉昭民, the evaporator was first made by John Leslei (1766-1832) in 1813.[9] Based on analysis of the manuscript, there seems little doubt that the rain gauge and evaporator described by Dai Yuan were of a Chinese type uninfluenced by Western design.

References

1. Joseph Needham, Science and Civilisation in China (Cambridge University Press, Cambridge, 1959) section 21, volume 3.
2. Wang Peng Fei 王鵬飛, 'Kai Chuang Zhong-Guo Qi-Xiang Shi Yan-Jiu de Xin-Ju-Mian' 〈開創中國氣象史研究的新局面〉 ('Exploring a New Approach to Research on the History of Meteorology in China') in Nanjing Qi-Xiang Xue-Yuan Xue-Bao 《南京氣象學院學報》(Journal of the Nanjing Meteorological Institute), 2, 129-138 (1984).
3. Guang-Ming Ri-Bao 《光明日報》(Guang-Ming Daily Newspaper) June 23, 1972; Li Di 李迪, 'Zhong-Guo Gu-Dai Guan Yu Qi-Xiang Yi-Qi de Fa-Ming' 《中國古代關于氣象儀器的發明》('On the Invention of Meteorological Instruments in Ancient China') in Da-Qi Ke-Xue 《大氣科學》(Atmospheric Science) 2, No. 1 (1978).
4. See also Fu-Hou Gu-Jin-Zhu 《伏侯古今注》(Commentary of the Lord Fu on Things New and Old) of 140 by Fu Wu-Ji 伏無忌 and Jing-Zhai Gu-Jin-Tou 《敬齊古今黈》(The Commentary of Jing-Zhai on Things Old and New) of the Jin-Yuan 金元 period by Li Ye 李治.
5. Wada Yuji 和田雄治, Seisō Ei-so Ryōchō no Soku-u-ki 〈世宗英祖兩朝之測雨器〉 'Korean Rain Gauges from the Reign of Sejong (1419 to 1450) and Yongjo (1725 to 1776)' Japanese Meteorological Journal, 3 (1911).
6. See Zhong-Guo Gu-Dai Du-Liang-Heng Tu Ji 《中國古代度量衡圖集》(Ancient Chinese Meteorology Illustrations) compiled by the Department of Measurements in China (Wen Wu 文物 Publishing Co., Beijing, 1981).
7. The date is consistent with Dai Yuan's work in music entitled Chun Cao Tang Qin Pu 《春草堂琴譜》(Chun-Cao-Tang Manual

for the Lute). It is a book revised in 1744 by Dai Yuan, Su Jing 蘇景 and Cao Shang-Jiang 曹尙裝 .

8. Li Yan 李儼 , Zhong Suan Shi Lun Cong《中算史論叢》(Collected Essays on the History of Chinese Mathematics) (Science Publ. House, Beijing, 1954), volume 2, p. 283.

9. Liu Zhao-Min 劉昭民 , Xi-Yang Qi-Xiang-Xue Shi《西洋氣象學史》(A History of Western Meteorology) (Taipei, 1981), pp. 93-94, 116.

SOME REMARKS ON THE CHINESE HISTORICAL EARTHQUAKE RECORDS

Shu-Qin Qi 齊書勤

China forms part of one of the world's greatest area of seismic disturbance. Based on extant records, 49 earthquakes of greater than 7th magnitude occurred in China before 1900, eleven of which were greater than 8th magnitude. It is therefore natural that the Chinese should extend their record keeping to include earthquakes. This frequent seismic activity and extensive historical records provide an exceptional advantage for studies in seismology. The recent publication of Zhong Guo Di Zhen Zi Liao Nian Biao《中國地震資料年表》(A Chronological Table of China's Earthquake Materials) in 1956 and Zhong Guo Di Zhen Li Shi Zi Liao Hui Bien《中國地震歷史資料彙編》(A Collection of China's Historical Earthquake Records) in 1983 has made available a valuable source of information for the systematic study of seismic activity in China. The following are some of the author's observations on this subject presented here with the hope of arousing further discussion.

I. The Role of Local Historical Documents

China's historical records of earthquakes are scattered through official histories, unofficial histories, biographical notes, chronicle books, miscellaneous records, patriarchal clan pedigrees, Buddhist and Taoist scriptures, tablet inscriptions, dedications, poems, local chronicles and locally stored documents. As far as their value as seismological records is concerned, the local documents, which often contain first-hand information, are sometimes of greater scientific value than records contained in the official documents which are polished second-hand material. However, many of such local documents from before the Yuan 元 dynasty (1271 to 1368) are no longer available. As a result, earthquake information from before the Yuan dynasty comes primarily from official documents based on reports to the throne by local officers on conditions of earthquake regions and imperial edicts on such matters as earthquake relief work and its evaluation.

After the Song 宋 dynasty (960 to 1279), local documents become available. They have played an important role in helping us to learn about earthquake activity in China. As an example, let us consider the earthquake which took place in the central and northern part of Shanxi 山西 province on September 17, 1303. It is not only recorded in the Yuan Shi 《元史》(The History of the Yuan Dynasty) but also in several local chronicles and tablet inscriptions (found through archeological studies) as well as dedications and other historical relics. Based on these records, we have ample evidence to claim that this earthquake was of the 8th magnitude. This is the first well documented 8th magnitude earthquake in China's history.[1]

With the help of local documents, as many as eleven 8th magnitude earthquakes have been determined from historical documents during the 600 years or so between 1303 and 1900. Before 1303, China was not without severe earthquakes, it is the lack of actual first-hand accounts that prevents us from identifying them from the official records. Therefore, in studying China's seismological history, one should, in addition to research into the official historical records, make use of the records and documents stored in local districts.

II. The Development of Seismology in the Han 漢 Dynasty

China's seismological history shows that the progress and development of seismology is, as expected, closely related to periods of high seismic activity. The occurrence of strong earthquakes would bring heavy casualities and heighten awareness of the various phenomena that are associated with earthquakes. The first sustained active period of earthquakes recorded in Chinese history occurred primarily along the north-south seismic belt, in the Fen 汾 and Wei 渭 river valleys and in the Tanlu 郯盧 seismic belt during the middle of the Han 漢 dynasty (-206 to 220). According to incomplete statistics there are more than 270 instances of earthquake recordings in the Shi Ji 《史記》 (Historical Records), the Han Shu 《漢書》 (History of the Han Dynasty) and the Hou Han Shu 《後漢書》 (History of the Later Han Dynasty) and other historical documents, totalling nearly 40,000 words altogether. Among these earthquakes, 106 were of medium magnitude, 23 above 5th magnitude and 6 above the 7th magnitude.

It was during this period that seismology in China developed rapidly. We saw the invention of the world's first seismograph in +132 by Zhang Heng 張衡, a well-known scientist who made significant contributions to astronomy and mathematics in addition to being the first inventor of a seismograph.[2] According to the records, the seismograph successfully recorded the earthquake in Lungxi 隴西, about 400 miles away from the capital, on February 28, +138 and was confirmed several days later by a message sent from Lungxi. The basic text on Zhang Hen's seismograph is contained in the biographical chapter devoted to him in the Hou Han Shu. There are a number of translations and paraphrases of the relevant passage in Western languages. A detailed exposition of Zhang Hen's seismograph is given by Needham in section 24 of his Science and Civilisation in

China[3] based on reconstructions of the seismograph's mechanism by Wang Zhen-To 王振鐸[4] and Imamura.[5] It is likely that Zhang Heng's seismograph was operated on the principle of an inverted pendulum. The apparatus was designed to determine the azimuth direction in which the earthquake's epicenter lay.

No great progress on the theory of earthquakes was made in China, as in all parts of the world, until the conception of the earth's crust in modern times. Earthquakes in ancient China were explained in terms of 'qi' 氣 and the 'yin-yang' 陰陽 theory. The invention of the seismograph by Zhang Heng placed a greater emphasis on the observation of the physical phenomena of an earthquake rather than on metaphysical interpretations of earthquakes. It was during this period that the term 'di zhen' 地震 was coined to replace 'di dong' 地動 for earthquake.[6] 'Di dong' means simply 'earth-movement' while 'di zhen' means earth movement in resonance with or in reaction to the source of movement. This change of a single word from 'dong' 動 to 'zhen' 震, was by no means meaningless. It reflected a better intuitive understanding of earthquakes since we now know that the earthquake that we feel is the local reaction to the earth's crust to the propagation of seismic waves radiating away from a distant seismic event.

III. Records of Unusual Phenomena Associated with Earthquakes

The Chinese earthquake records often contain information on unusual occurrences of natural phenomena and animal behaviors prior to, during and after earthquakes. Although the connection between such occurrences and earthquakes is not always apparent, this information is nevertheless of importance since it might provide clues for the understanding or prediction of earthquakes.

Keeping records of such unusual occurrences began very early in China. One of the earliest extant earthquake records is about an earthquake that took place near Yongji 永濟 in Shanxi 山西 province in the -23rd century. The record was undoubtedly transmitted orally for generations before it was recorded in Chou books such as the Mo Zi 《墨子》 (The Book of Master Mo). We have for the Yongji earthquake the description from volume 5 of the Mo Zi:

> 日妖宵出，雨血三朝。龍生廟，犬哭乎市。
>
> 夏冰，地坼及泉，五谷變化。
>
> The sun appears unnatural as darkness sets in with blood-colored rain falling for three days. Snakes appear in temples, dogs bark in streets. Water freezes in the summer, the earth splits to form springs and crops (all five grains) grow untimely.

This record provides information on weather, animal behavior, and ground movement as well as the effects on agricultural products.

About the earthquake in Lungxi in -193, the Han Shu gives the following account:

> On the 4th day of the first month, two snakes appeared in the well of Lanling 蘭陵 but two days later they disappeared at dusk. Not long after Lungxi 隴西 experienced a violent earthquake.

The account apparently implies that there is a connection between the abnormal appearance of snakes in cold weather and earthquakes.

Such records are often dismissed as superstitious since the occurrence of two events may have been coincidental. However, an animal's senses of perception are different from those of humans. Recent research in China indicates that there are certain correlations between abnormal animal behaviors and the imminent occurrence of an earthquake. In this respect, studies of past records of unusual phenomena may be useful.

IV. Relief and Prevention of Earthquake Disasters

Ancient Chinese records of earthquakes also contain information on earthquake relief and the prevention of earthquake disasters. The first instance of the social impact of earthquakes being noted was on June 1st, -70, when a 7th magnitude earthquake occurred in Zhucheng 諸誠 and Changle 昌樂 counties of Shandong 山東 province, which brought about such severe calamities that Emperor Xuan 宣 of the Han 漢 dynasty decreed exemption from taxation and rents and a general pardon throughout the county. On October 13th, -67, he further decreed:

> Those who are happily alive, if poor, may live in public houses without building their own houses; if returned to their homeland, may cultivate state land and the state will plant seed for them to eat.[7]

These measures were taken to rehabilitate and restore the economy. This was the first policy providing disaster relief recorded in China's history.

Experiences accumulated from earthquakes led also to anti-earthquake projects which encouraged homeowners to reinforce doorways with wooden frames and ceilings with wooden beams. People were instructed to stay calm instead of running around during an earthquake. In the passage concerning the earthquake which took place in 1679 at Sanhe 三河, Beijing 北京, we read the observation that in most cases by hiding under windows, or tables, or behind doors, one can avoid being killed by earthquakes. Books on earthquakes with precaution instructions began to appear.[8]

V. Earthquake Records and Seismic Research

The distribution in location and frequency of earthquake is closely related to the fracture structure and its activity in the earth's crust. The north-south, and west-east geological structure zones divide most of China into five seismic regions with different geologic structural characteristics and seismic activity levels. The information provided by historical earthquake records is of importance not only to the identification and confirmation of these seismic regions but also to the determination of the periodicity of their seismic activity.

Historically, the earthquakes in West China mainly take place along the sharply swelled area surrounding the Qinghai-Xizang 青海西藏 plateau, the Hengduan 横断 mountain range, on both sides of Tian-shan 天山 mountain, and in the region of the Qilianshan 祁連山 mountains. The earthquakes in the eastern portion of China are mostly concentrated on the sharply sunken plains, fracture-sunken basins, or in the vicinity of large fracture zones which are active in modern times, such as the Fen-Wei 汾渭 graben, the Hebei 河北 plain, and the big fracture zone along the Tancheng 郯城 and the Lujiang 盧江 regions. All these earthquakes are distributed in strips and occur densely in definite regions of geological structure. In addition, more than 80 percent of the earthquakes of the 7th magnitude or above took place in the comparatively large fracture zones. This close association is also found in other seismic regions. In the Shanxi 山西 seismic zone, earthquakes of the 6th magnitude and above have all taken place in the Cenozoic era fracture-sunken basins.

The new fractures of the earth's surface caused by strong historical earthquakes are also in conformity with the main fracture directions, sometimes even overlapping each other. This is supported, for example, by the 8th magnitude earthquake which took place in Hongtong 洪洞, Shanxi 山西 province in 1303, the 8.5 magnitude earthquake in Haiyuan 海原, Ningxia 寧夏 region in 1920, the 8th magnitude earthquake in Fuyun 富蘊, Xingiang 新彊 region in 1931, the 7.5 magnitude earthquake in Dulan 都蘭, Qianghai 青海 province in 1937 and a number of other earthquakes of over 7th magnitude.

Analyzed from the vertical direction, the centers of shallow earthquakes in China's history are generally at a depth of 50 km from the surface of the earth. The depth of earthquake sources in the Shanxi seismic zone are shallower still, about 10 km from the surface. From this it can be seen that the earthquake sources are mostly in the crystal base of the earth's crust or even near its top surface. The great fracture zone along which frequent strong earthquakes took place in China's history often manifest a downward fracture to the depth of the earthquake source.

In studying earthquakes, it is instructive to learn the relationship between the vertical differential motion and the horizontal motion. All of the strong earthquake regions are accompanied by marks of the vertical differential motion of the Cenozoic era, and all are closely related to the sharp sunken basins of the Cenozoic era. The 7 1/4-magnitude earthquake which took place in Dingxiang 定襄 county, Shanxi 山西 province in 1038 and the 8th magnitude earthquake in Hua 華 county, Shaanxi 陝西 province in 1556, for example, are both connected with the regions of the sharpest vertical differential motion during the Cenozoic era and afterwards. Such examples are also common in the historical seismic records of other strong earthquake regions.

China's earthquake records also provide valuable information on the pattern of energy accumulation and release in the pregnancy and occurrence of earthquakes. In the North China seismic region, for example, statistical information on earthquakes is quite comprehensive after the 10th century. This permits us to differentiate four major periods of earthquake activity during the past 1000 years, as summarized in Table I.

Table I. Periods of Earthquake Activity in the North China Seismic Region

Major Periods	Active Interval	Inactive Interval
1011 - 1076	65	214
1290 - 1368	78	116
1484 - 1730	246	82
1812 -	> 174	-

The Shanxi earthquake zone is another region where there are sufficient statistical records to provide information on local periodic changes in time. The historical earthquakes which took place in Shanxi before 1200 can be divided into different stages and periods according to earthquake frequency and strength. For earthquake activity since 1200, Shanxi can be divided into three periods which are roughly in conformity with the periods of large scale activity in North China.

The Chinese earthquake records have a range of applications in seismic research. Recent efforts in the systematization of these records will undoudtedly enhance their accessibility and usefulness not only in Chinese research but also in global seismic research.

References

1. See for example, Zhoug-Guo Di-Zhen Mu-Lu 《中國地震目錄》 (A List of China's Earthquakes) (Science Publ. Co., Beijing, 1983), edited by Gu Gong-Xu 顧功敍.
2. Qi Shu-Qin 齊書勤, 'The Active Period of Earthquakes in the Han Dynasty and Zhang Heng's Invention of the Seismograph' in the Proceedings of the Second National Symposium on the History of Earthquake Science (1984).
3. Joseph Needham, Science and Civilisation in China (Cambridge University Press, Cambridge, 1959), Volume III, section 24, pp. 624-635.
4. Wang Zhen-To 王振鐸, 'Han Zhang-Heng Hou Feng Di-Dong' Yi Zao-Fa Zhi Tui-Ce' 〈漢張衡候風地動儀造法之推測〉 ('A Conjecture as to the Construction of the Seismograph of Zhang Heng of the Han Dynasty') in Yenching Xue-Bao 《燕京學報》 Yenching University Journal of Chinese Studies, 20, 577 (1936); 'Di Dong Yi' 〈地動儀〉 ('Zhang Heng's Seismograph') in Ren Min Hua Bao 《人民畫報》 (People's Illustrated) April (1952).
5. Imamura Akitsune 今村明恒, '1800-Nen mae no Jishinkei' 〈千八百年前の地震計〉 ('A Seismograph of Eighteen Hundred Years Ago') in Shina Bunka Dansō 《支那文化談叢》 (Tokyo, 1942).
6. The term 'di zhen' 地震 first appeared in records of the -193 earthquake in Lintao 臨洮, Gansu 甘肅 in both the Han Shu and the Qian Han Ji 《前漢記》 (Records of the Former Han Dynasty).

7. See the biographical section of Emperor Xuan 宣 in volume 8 of the Han Shu 《漢書》 (History of the Han Dynasty).

8. Qin Ke-Da 秦可大, Di Zhen Ji 《地震記》 (Earthquake Records), 1575.

TECHNOLOGY

THE DEVELOPMENT OF METAL TECHNOLOGY IN ANCIENT CHINA

Tsun Ko 柯俊

The use of metals and the evolution of metallurgy are basic to the progress of society and civilization. Metallurgical technique has provided copper, iron and their alloys which has in turn furnished and are still furnishing implements, weapons, and constructional materials. The increase in productive forces and productivity associated with the use and improvement of metals and alloys has provided the active power and material basis for social progress. The development of ancient civilizations in Mesopotamia, the Nile Valley, the Indus Valley and China, and even later and modern civilizations has in one way or another been related to the use and the science of metals.

I. The Bronze Age in China

Chronologically, metals came into use later in Chinese civilization than in other ancient civilizations.

The earliest metal artifact found in China so far is a bronze knife of the Majiayao 馬家窰 culture, excavated at Dongxiang 東鄉, near Lanzhou 蘭州, Gansu 甘肅 province, near the upper reaches of the Yellow River and the Wei 渭 River, which passes present day Xian 西安. It contains 5-10% tin as estimated from the microstructure, and dates from -2800 to -3000. The fact that the date corresponds to that of early bronzes in Ur remains supports the view that the development of metallurgy in China was indigenous.

China entered the bronze age around the -22nd century. Numerous artifacts of copper and copper alloys, including a brass awl (in two pieces), copper-tin, copper-lead and copper-tin-lead bronze tools, bronze, gold and silver ornaments, and mirrors of the period between the -21st and the -16th centuries have been excavated in the Yellow River basin, mostly in Gansu, where copper ore is abundant. Recently, a fragment of a bronze vessel of the -21st century, believed to be of the early Xia 夏 dynasty, was found in Dengfeng 登封, Henan 河南 (Figure 1).[1] A stone mold for casting arrowheads

Figure 1. A fragment of a bronze vessel of the -21st century unearthed at Dengfeng 登封, Henan 河南 (see Ref. 1).

from earlier than the -16th century was unearthed in Gansu. Thirty bronze articles including bells, awls, and a 'tong jue' 銅爵 ('bronze ceremonial wine vessel') (Figure 2) of the -17th century, and about 30 bronzes of the -16 century have been found in Erlitou 二里頭, Henan believed to have been one of the Shang 商 capitals.[2] Two square 'tong ding' 銅鼎 ('bronze cauldrons') weighing 82 kg (Figure 3) and 64 kg, respectively, of the mid-Shang dynasty were excavated in Zhengzhou 鄭州, another Shang capital, the former

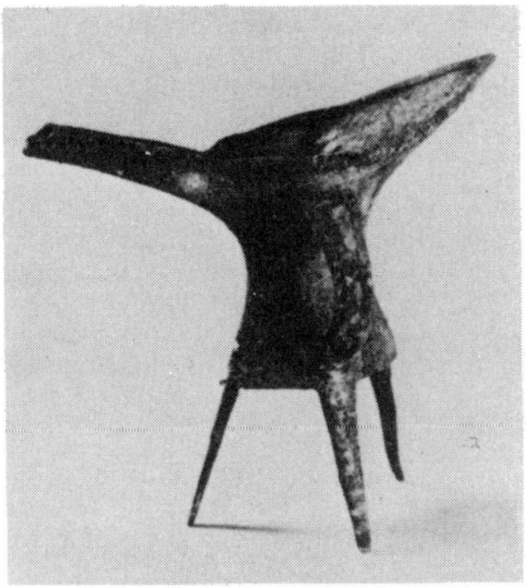

Figure 2. A bronze 'jue' 爵 ('ceremonial wine vessel') of -16th century unearthed at Erlitou 二里頭, Yanshi 偃師, Henan 河南 (see Ref. 2).

Figure 3. A bronze 'ding' 鼎 ('cauldron') of c. -15th century (weight 82.4 kg, height 100 cm, mouth length 62.5 cm, mouth width 60.8 cm and depth 46 cm) unearthed in 1974 at Zhengzhou 鄭州, Henan 河南.

containing 17% lead and 3.5% tin. The 'si-mu-wu' 司母戊 'ding' (Figure 4) of the late -13th century containing 85% copper, 12% tin and 2.8% lead, weighing 875 kg and cast in a composite clay mold, is the largest ancient casting of the -2nd millennium found to date in the world. Large numbers of bronze articles and foundry sites from the Zhou 周 dynasty have been found, along with thousands of mold pieces, including positive mold pieces for making impressions on the molds. Most of these castings were made using composite clay molds, the lost-wax method came into use only in the -6th century, but it had been developed into a sophisticated technique for intricate castings by the -5th century.

For melting and smelting, pottery cisterns with conical bottom covered with clay inside and clay mixed with straw outside were used in the early stages of development. Charcoal and copper or copper

Figure 4. A large bronze 'ding' 鼎 ('cauldron') of the late -13th century (weight 875 kg, height 133 cm, length 110 cm, and width 78 cm) engraved with the three characters 'si mu wu' 司母戊 unearthed at Anyang 安陽, Henan 河南.

ore were placed inside the crucible. The straw-mixed clay covering served as an insulating material. This was developed into a shaft furnace constructed from pottery cylinders and faced with clay on both sides, with openings provided for air supply and tapping. Found buried in the 400,000 tons of copper smelting slag in Tonglu-Shan 銅綠山 (i.e. Mt. Verdigris), Daye 大冶 (or, 'Grand Smeltery'), Hubei 湖北, an ancient copper mine worked between the -10th century and the +2nd century,[3] the smelting furnaces had developed into shaft furnaces with a dome-shaped top opening and a hollow underneath the hearth to keep away the dampness.

Early records show that by the -3rd century, enough experience on the relation between the properties and composition of bronzes had been accumulated that six different tin contents were given for uses which required different properties.[4] This is the earliest known rule of alloying. It is also stated in the Lu-Shi Chun-Qiu 《呂氏春秋》 (Master Lu's Spring-Autumn Annals) of -239 that "mixing the soft metals copper and tin produces a hard material," and in the Xun Zi 《荀子》 (The Book of Master Xun) by Xun Kuang 荀況 (b. -313, d. -238) that when a casting was to be made, it was important to have a properly made mold, copper and tin of high quality, proper and refined smelting, correct composition, and correct heat and temperature.

II. The Iron Age

Meteoritic iron was used in making ritual weapons dating to between the -14th and -12th centuries. China entered the iron age by the middle of the -7th century. The earliest bloomery iron artifacts excavated include a fragment of an iron sword of the mid-seventh century from a Qin 秦 tomb at Lintai 靈臺, Gansu 甘肅, about 150 km from Xian in the upper reaches of the Jing 涇 River, an iron rod found in a tomb of the late -6th century (Figure 5), a knife and a sword of the same period found in Changsha 長沙, Hunan 湖南. The structure of bloomery iron as exemplified by the -6th century iron rod is given in Figure 6. Bloomery iron and steel made from it by cementation were in use until the late -2nd century when it was replaced by steel made from pig iron by decarburization in the solid state (oxidation annealing) or in the liquid state (puddling).

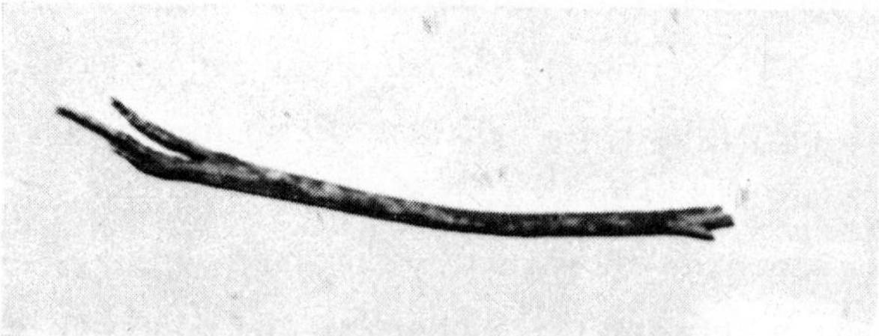

Figure 5. A portion of an iron rod (25.5 cm long) of the late -6th century unearthed in 1974 at Chengqiao 程橋 in Liuhe 六合, Jiangsu 江蘇.

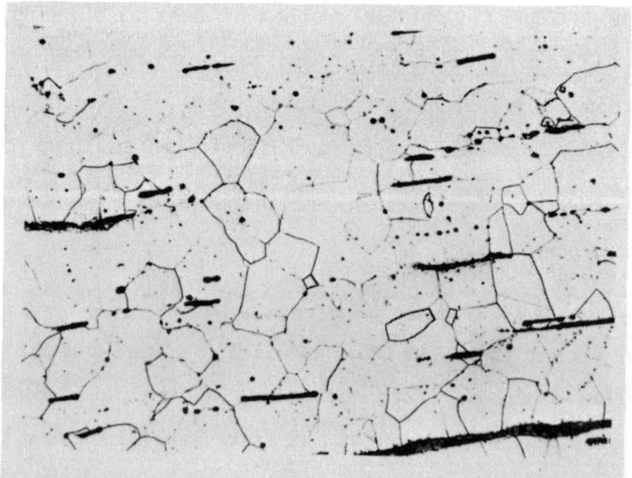

Figure 6. The structure (enlarged 150 times) of bloomery iron as exemplified by the iron rod of the -6th century shown in Figure 5.

A significant contribution of ancient Chinese metallury was the use of cast iron, which became widespread due to the invention of the malleabilizing process of white iron, and the making of steel from cast or pig iron. This evolution began in the -6th century and was almost completed by the +2nd century. The production of iron in a liquid state, the making of malleable iron from castings, and the making of wrought iron and steel from pig or cast iron greatly facilitated their production and increased productivity, with increased supply and lowered cost. The development played a very important part in the advance of agriculture and water irrigation, in strengthening the national economy and military power, and was one of the significant factors in the formation of a unified China and in the evolution and continuation of Chinese civilization.

The production of cast iron was the natural consequence of the high temperatures obtained by the early kiln technique, which also led to the invention of porcelain ware in China, in spite of the fact that China clay, i.e. Kaolin 高嶺 clay, is available in many places in the world. The direct heating of copper and copper alloys in contact with burning charcoal, and in many cases the use of baked hot clay molds account for the fine details obtained in ancient Chinese castings. The high temperature achieved in the iron smelting furnace caused the absorption of carbon from the fuel into the iron, lowering its melting point, thus further facilitating the carbon absorption, and eventually produced liquid iron with 3 - 4% carbon and a melting point as low as 1130°C.

Cast iron objects of the -6th century have been found in Changsha 長沙 and near Nanjing 南京 in the Yantze 揚子 valley. An adze made of white iron dating back to the early -5th century has been shown to contain a low carbon ferritic layer on the exposed surfaces as a consequence of annealing in a decarburizing atmosphere.

Annealed cast iron with a structure corresponding to modern malleable iron was also found in a spade excavated together with the abovementioned adze. It is evident that the invention of malleable iron soon after white iron castings were produced was the consequence of an attempt to soften the white iron by annealing. This invention provided a cheap method for producing cast iron implements with a certain toughness in large quantity, and enabled the production of large ploughshares for deep furrowing, and sharp and strong but inexpensive sickles for harvesting the crops, the yields of which were at that time increasing. The spread of iron farming implements in the -4th century, as evidenced from archeological findings, can also be inferred from a recorded discussion by Mencius 孟子 (b.c.-390, d.-305), who asked his student Chen Xiang 陳相 "Is Xu Zi 許子 farming with iron?,"[5] the first reliable mention of iron in ancient Chinese literature.

The production of iron castings and malleable iron was furthered by the invention and use of permanent metal molds (Figures 7 and 8). By the Zhan-Guo 戰國 period, the 'die zhu' 疊鑄

Figure 7. Permanent cast-iron of the -4th century for casting an adze unearthed in 1956 at Xinglong 興隆, Hebei 河北.

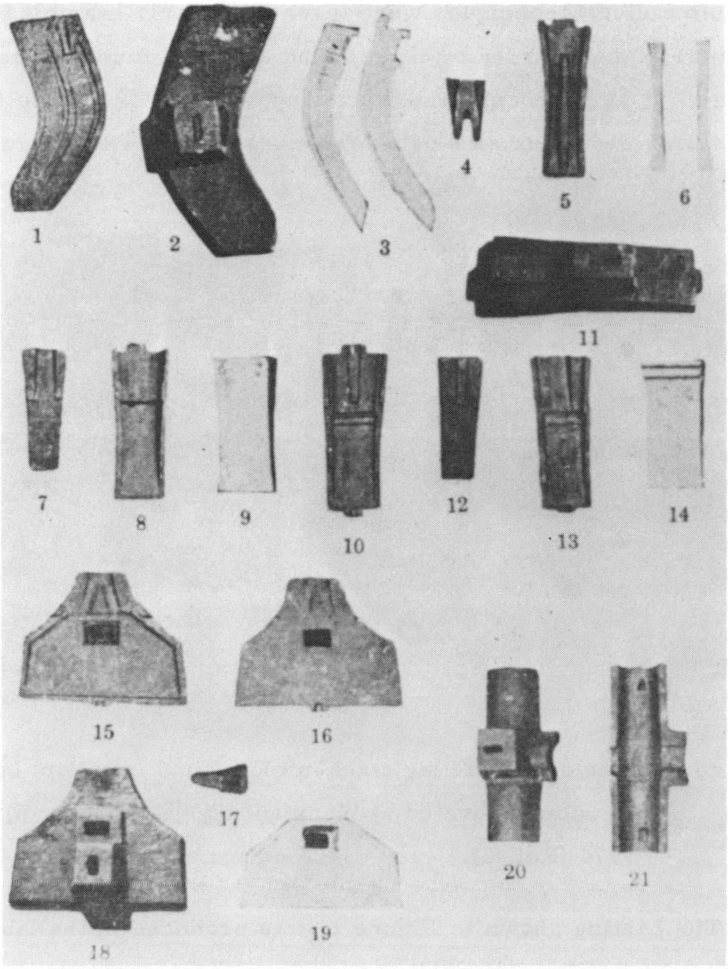

Figure 8. Iron-casting molds and cores of the -4th century unearthed at Xinglong, Heipei. 1-2: Mold for two sickles. 3: Sickles cast from the mold shown in 1-2. 4: Core for mold numbered 5. 5: Mold for two chisels. 6: Tin chisels cast with mold 5. 7-8: Adze mold and core. 9: Tin adze cast in molds 7 and 8. 10-11: Mold pieces for an axe. 12: Core for axe mold. 13: Mold assembly for axe. 14: Tin axe cast in mold 11. 15-16: Mold pieces for hoe. 17: Core insert. 18: Mold assembly for hoe. 19: Tin hoe cast with mold 18. 20-21: Mold pieces for carriage parts.

('stack-molding') techniques were invented. Early examples of multi-layer molds are afforded by those found at an old foundry site in Wenxian 温縣, Henan. The discovery of a Han 漢 baking furnace in 1974 provided many sets of stack-molds[6] as shown in Figures 9 and 10.

Figure 9. Sets of baked clay stack-molds found in a Han 漢 baking furnace excavated at Wenxian 温縣, Henan 河南 in 1974 (Ref. 6).

The casting shown in Figure 11 was produced in the laboratories of the Beijing University of Iron and Steel Technology (BUIST) using baked clay stack-molds found in the above mentioned Han 漢 baking furnace.[7]

The wide application of cast iron promoted iron production. The hearths of two blast furnaces of the Han dynasty with major and minor axis measuring 4 m and 2.7 m across, respectively, together with salamanders weighing 20-25 tons buried in front of the furnaces and blocks of furnace charges fused together have been uncovered in Guxing 古滎 near Zhengzhou, Henan (Figure 12).

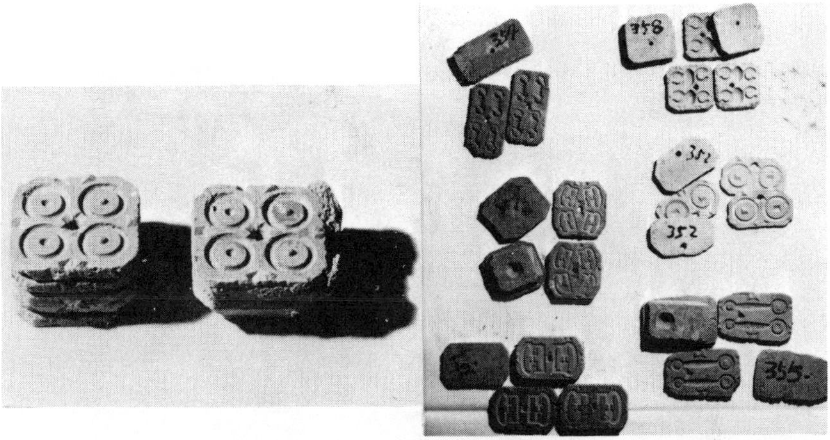

Figure 10. Disks from disassembled, baked clay stack-molds found in a Han 漢 baking furnace excavated at Wenxian 温縣, Henan 河南 (Ref. 5).

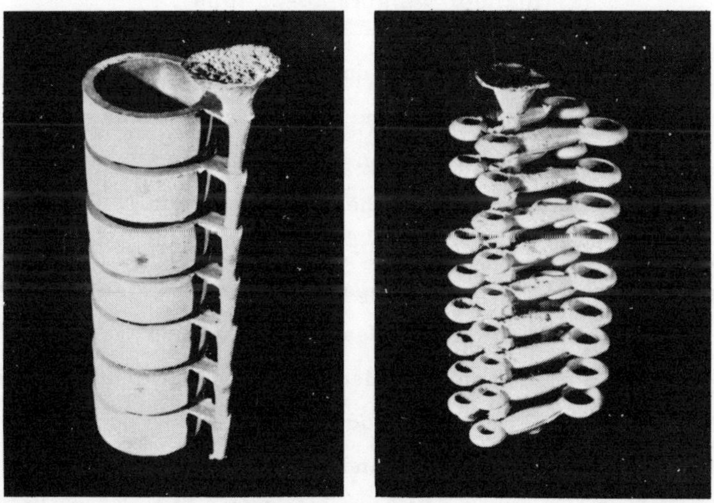

Figure 11. Castings produced using the baked clay stack-molds found in a Han 漢 baking furnace excavated at Wenxian 温縣, Henan 河南 (Ref. 7).

Figure 12. A salamander from blast-furnaces of the Han dynasty excavated in Guxing 古滎 near Zhangzhou 鄭州, Henan 河南 (major axis 4 m, minor axis 2.7 m, salamander weights 20-25 tons).

Two methods of steel production from pig iron were invented in the late -2nd century, one by the decarburization of cast iron in the solid state without the formation of graphite and subsequently used either in the cast shape (e.g. the scissors found in Dongshima 東史馬, Zhengzhou, Figure 13)[8] or forged, the other by puddling of molten iron. The former is regarded as a new steelmaking process distinct from black hearth malleable iron without graphite, since cast iron plates and corresponding decarburized material containing no graphite (steel) and articles produced therefrom by forging have been found at a site of ancient iron and steel works. The products of both of these processes can be distinguished from bloomery iron by the composition, constituents and distribution of the inclusions and the segregation of trace elements in the matrix.

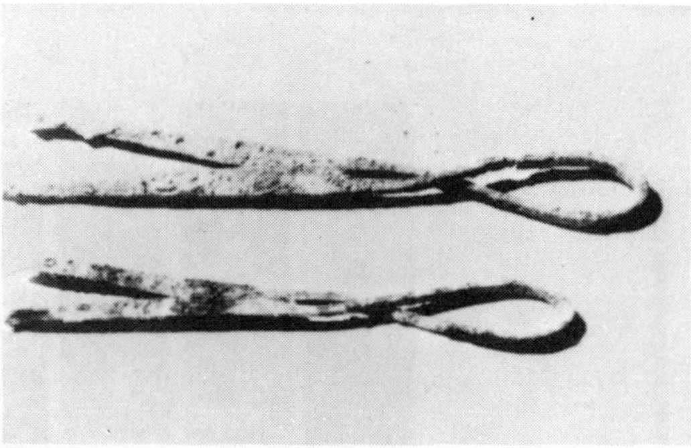

Figure 13. Two pairs of steel scissors made by decarburization of iron castings unearthed in Dongshima 東史馬, Zhengzhou 鄭州, Henan 河南 (Ref. 8).

Improvement of the steel quality by repeated forging, folding and stacking of wrought iron or steel made by the puddling process from pig iron was in practice by the +1st century. Metallographic study has shown that the steel was named in accordance with the number of layers in the steel after repeated forging. A single bladed sword of +112 unearthed in 1974 in Shangdong 山東 is engraved with the name 'sa-lian da-dao' 卅涑大刀 ('30-refinings large sword')[9] as shown in Figure 14. The layers in such forged steel can be readily studied. In Figure 15, a portion of the laminar structure of a 50-refinings sword of +77 is shown.[10] Knives made of 30-refinings steel dating back to between the +2nd and +3rd centuries and a sword of 100-refinings (found in Japan) dating back to between +185 and +186 are also known to exist.[11] The great difference between the iron and steel making in China and that in the other parts of the world is reflected in the composition of the ancient slags of other countries.

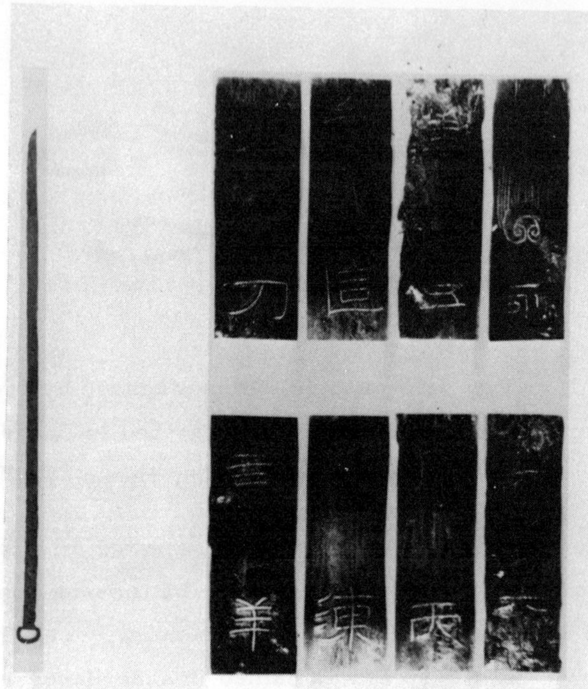

Figure 14. A steel sword (left) engraved with the name 'sa-lian da-dao' 卅湅大刀 ('30-refinings large sword') and the date 'Yong-Chu liu-nien wu-yue bing-wu' 永初六年五月丙午 ('made in the fifth month of the sixth year of the Yong-Chu reign,' i.e. +112) (right) unearthed in 1974 at Cangshan 蒼山, Shangdong 山東 (Ref. 9).

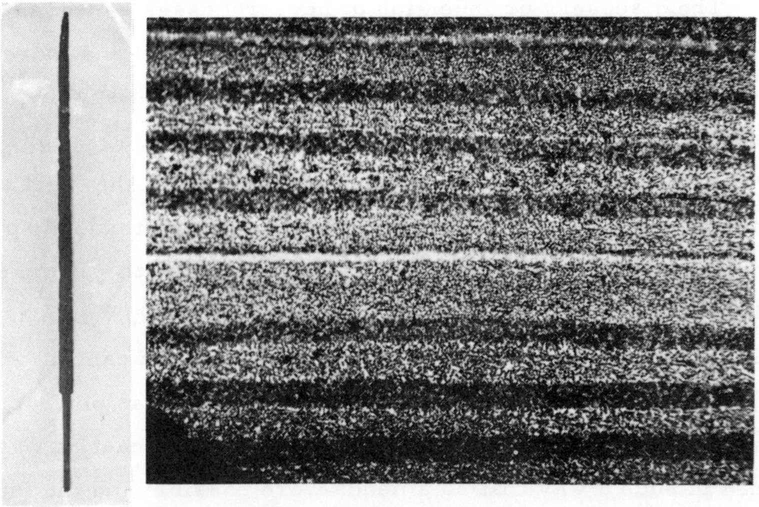

Figure 15. Layer structure of steel (right) of a 50-refinings steel sword of +77 (left) unearthed at Xuzhou 徐州, Jiangsu 江蘇 (Refs. 10 and 11).

III. Other Metallurgical Achievements in China

Other achievements in ancient Chinese metal production and fabrication included the invention of gilding with amalgam (-5th century), the use of copper-nickel alloys (4th century) and copper-nickel-zinc alloys, Paktong ('bai-tong' 白銅) or 'German silver,' (12th century), the extraction of copper from copper-bearing streams by ionic replacement with iron (first mentioned in the early -2nd century and in large scale production not later than the 12th century) and the use of silver-tin amalgams for dental filling of tooth decay in the Tang 唐 dynasty and first mentioned in the Imperial Pharmacopoeia of 659. It is stated that the alloy hardened on standing, probably the earliest record or discussion of the age-hardening of an alloy. In smelting, coal was used in iron making in blast furnaces from the 10th century on.

The discovery or invention of new processes slowed down after the 12th century, but the development of technology and increase of production continued. The cannons, large Buddhist statues, and astronomical instruments, that were cast, the large anchors that were forged, and the export of zinc and Paktong to the West in the centuries that followed all bear witness to the craftmanship of this period, which reached its height in the casting, with a composite clay mold, of the Yongle 永樂 Bell in Beijing (c. 1420), weighing 46-1/2 tons and covered both outside and inside and on the cannon with 227,000 characters in relief. The craftmanship was perfect. The bronzesmiths were so sure of themselves that the casting was designed in such a way that no amends were possible and the founding so well executed that no repair could be found - the evidence of a workmanship achieved through the accumulation of expertise over three millenniums.

Acknowledgments: The above account is a very brief summary of part of the work carried out by the BUIST Archaeometallurgy Group in collaboration with archeologists, museum curators, metallurgists and chemists in various laboratories in the whole country. The author is indebted to them for their cooperation and to his colleagues, Professors Qiu Liang-Hui 邱亮輝, Han Ru-Bin 韓汝玢, Huang Wu-Di 黃務滌 and many others in the Group for their constant support. The author can hardly express his gratitude to Professor Joseph C. Y. Chen 程貞一 for his encouragement and patience.

References

1. 'Dengfeng Wangchenggan Yi-Zhi de Fa Jue'〈登封王城崗遺址的發掘〉('The Excavation of the Ancient Sites at Wangcheng-gan in Dengfeng') by the Henan 河南 Archeological Research Institute and the Archeological Section of the Museum of Chinese History in Wen Wu《文物》(Cultural Relics) 3, 8-20 (1983).

2. See the report of the Erlitou 二里頭 team of the Archeological Institute, Academia Sinica, 'Yanshi Erlitou Yi-Zhi Xin Fa-Xian di Tong-Qi he Yu Qi'〈偃師二里頭遺址新發現的銅器和玉器〉('New Discoveries of Bronze and Jade Artifacts at Yanshi, Erlitou Site') in Kao Ku《考古》(Archeology) 4, 259-263 (1976) and the report of the Luoyang 洛陽 excavation team of the Archeological Institute, Academia Sinica, 'Henan Yanshi Erlitou Yi-Zhi Fa-Jue Jian-Bao'〈河南偃師二里頭遺址發掘簡報〉('A Brief Report on the Excavation of Erlitou Site in Yanshi, Henan') in Kao Ku《考古》(Archeology) 5, 302-309 (1975). See also the report '1980 Nian Qiu Henan Yanshi Erlitou Yi-Zhi Fa-Jue Jian-Bao'〈1980年秋河南偃師二里頭遺址發掘簡報〉('Report on the 1980 Autumn Excavation of a Shang Site at Erlitou in Yanshi, Henan') by the Erlitou Archeological Team, IA, CASS in Kao Gu《考古》(Archeology) 3, 199-205 (1983).

3. Xia Nai 夏鼐 and Yin Bao-Zhang 殷寶璋, 'Hupei Tonglu-Shan Gu Tong-Kuang'〈湖北銅綠山古銅礦〉('An Ancient Copper Mine in the Tonglu Mountains of Hubei'), Kao-Gu Xue Bao《考古學報》(Journal of Archeology) 1, 1-14 (1982) and see also the report 'Hupei Tonglu-Shan Chun-Qiu Zhan-Guo Gu Kuang-Jing Yi-Zhi Fa-Jue Jian-Bao'〈湖北綠銅山春秋戰國古礦井遺址發掘簡報〉('A Brief Report on the Excavation

of the Ancient Copper Mine of the Chun-Qiu Zhan-Guo period in the Tonglu Mountains of Hubei') in Wen Wu 《文物》(Cultural Relics) 2, 1-12 (1975).

4. See the Kao Gong Ji 《考工記》(The Artificers Record).

5. See 'Teng Wen Gong Shang Pian'〈滕文公上篇〉in Meng Zi 《孟子》(The Book of Master Meng).

6. 'Henan Wenxian Han-Dai Hong-Fan-Yao Fa-Jue Bao Gao'〈河南温縣漢代烘範窰發掘報告〉('Report on the Discovery and Excavation of a Han Baking Furnace in Wen County of Henan Province') by the Henan 河南 Museum in Wen Wu 《文物》(Cultural Relics) 9, 66-75 (1976).

7. Han-Dai Die-Zhu — Wenxian Hong-Fan-Yao de Fa-Jue he Yan-Jiu 《漢代疊鑄—温縣烘範窰的發掘和研究》(Stack-Molding Casting in Han Dynasty — The Excavation and a Study of the Mold-Baking Furnace at Wenxian) (Wen Wu 文物 Publishing House, Beijing, 1978) by the Henan 河南 Provincial Museum and the Editorial Board of the History of Ancient Chinese Metallurgy.

8. Han Ru-Bin 韓汝玢 and Yu Xiao-Xing 于曉興, 'Dong-Shima Jian-Dao yu Gu-Ti-Tuo-Tan-Gang'〈東史馬剪刀與固體脱碳鋼〉('The Dong-Shima Scissors and the Steel-Making by Solid-State Decarburization of Cast Iron in Ancient China') in Zhong-Yuan Wen Wu 《中原文物》(Cultural Relics from the Central Plains) (Special Issue) 239-241 (1983).

9. 'Shandong Cangshan Fa-Xian Dong-Han Yong-Chu Ji-Nian Tie-Dao'〈山東蒼山發現東漢永初紀年鐵刀〉('An Iron Sword Engraved with a Date from the Yong-Chu Reign of the Eastern Han Discovered in Cangshan, Shandong'), Wen Wu 《文物》(Cultural Relics) 12, 61 (1974).

10. 'Xuzhou Fa-Xian Dong-Han Jian-Chu Er-Nian Wu-Shi-Lian Gang-Jian'〈徐州發現東漢建初二年五十湅鋼劍〉('The Steel Dagger Datable to the 2nd Year of the Jian-Chu Reign of the Eastern Han Unearthed in Xuzhou, Jiangsu') by the City Museum of Xuzhou in Wen Wu 《文物》 (Cultural Relics) 7, 51-52 (1979).

11. Han Ru-Bin 韓汝玢 and Ko Tsun 柯俊, 'Zhong-Guo Gu-Dai de Bai-Lian Gang'〈中國古代的百煉鋼〉('The Hundredfold Tempered Steel in China') in Zi-Ran Ke-Xue Shi Yan-Jiu 《自然科學史研究》 (Research in the History of Natural Science) 4, 316-320 (1984).

THE DEVELOPMENT OF TRIBOLOGY IN ANCIENT CHINA

Jing-Yan Lu 陸敬嚴

Tribology has appeared recently as a kind of independent science. Its appearance is by no means without precedent and its development can be traced back to remote antiquity. In this paper, the author gives an account of this development based on the examination of archeological finds and historical documents.

A preliminary summary of the accomplishments in tribology in ancient China having to do with the sliding bearing is presented. The events covered include the early sliding bearing, the appearance of lubrication, the early lubricants, lubrication methods and lubricating devices, the application of mineral oil in lubrication, the selection of wood for making sliding bearings, the application of metal axle bushings, the differences between the cart and the chariot in the length of the wheel boss, and replacing sliding friction by rolling friction. Through studying historical records, we can understand the development and change of tribology in ancient China. These records will also be useful to future research work.

I. The Early Sliding Bearing

The appearance of the sliding bearing made necessary the use of lubricants and fostered the development of tribology, Hence, we should know when the sliding bearing was used in ancient China and where it was applied. For this reason, in Table 1 are listed some machines which appeared early in ancient China to which sliding bearings were applied. In order to compare these machines, the dates of their appearances are also listed.

Table 1. Ancient Chinese Machines Based on the Sliding Bearing

Machine and Appearance Date	Designs	Sources
The cart: period between -2698 and -2599.	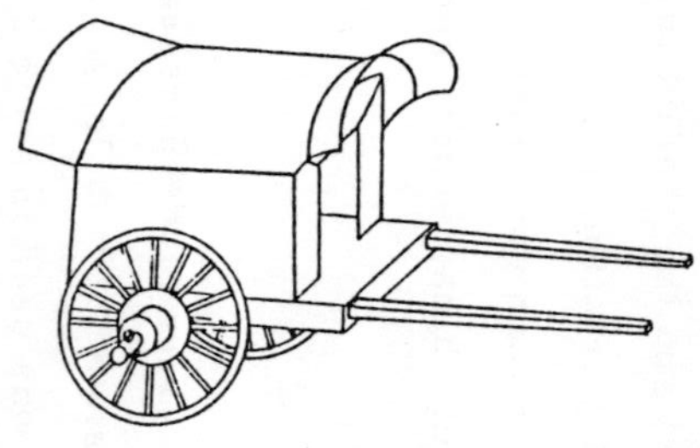	The Gu-Shi Kao 《古史考》(Examination of Ancient History) of the +3rd century states, "Huang-Di 黃帝 invented the cart." The design shown is an interpretation by an artist of the Northern Wei dynasty.
Treadle-operated tilt hammer: the period after the era of the Huang-Di 黃帝 (-2599), before the Yao Shun 堯舜 era (-2000). Water-operated tilt hammer: about -2000.	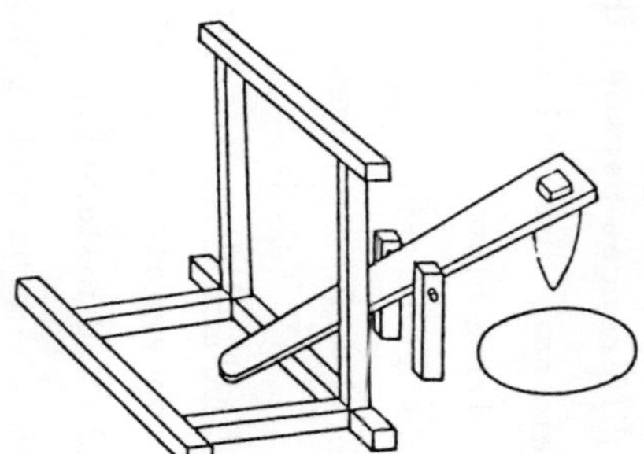	The book Xin-Lun 《新論》(New Discussion) of +20 states, "Fu Yi 伏羲 made the mortar and pestle." The book Wu-Yuan 《物原》(The Origin of Things) of the Ming dynasty states, "The water-operated tilt hammer was invented by Hou Ji 后稷. The design shown is a later illustration found in the Tian-Gong Kai-Wu 《天工開物》(The Exploitation of the Works of Nature) of 1637.

Table 1. Continued.

Machine and Appearance Date	Designs	Sources
The windlass: about -1100.		The book Wu-Yuan states, "Shi Yi 史佚 invented the windlass." The design shown is a later illustration found in the Tian-Gong Kai-Wu.
The pulley: before -800.		Found by archeologists at Zhangjiapo 張家坡, Xian 西安.[a]
The spinning wheel (operated by hand): before -200.	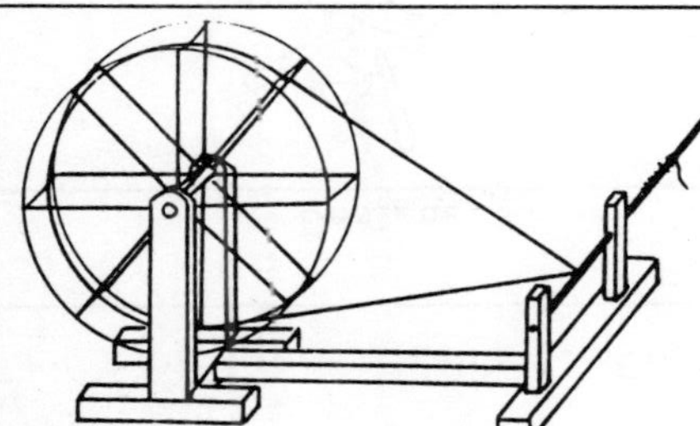	Found by archeologists in Tengxian 滕縣 Shandong 山東.[b]

Table 1. Continued.

Machine and Appearance Date	Designs	Sources
The winch: before the Western Han dynasty (-206 to +8).	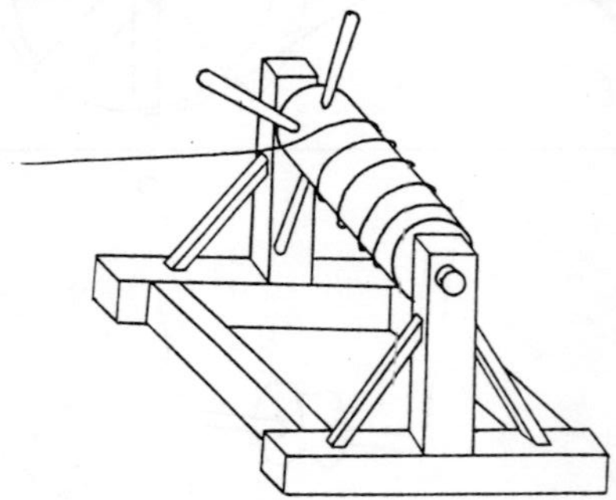	Found by archeologists at Tonglu-shan 銅綠山,[c] Hubei 湖北. The design shown is an illustration from the Xin-Zhi Zhu-Qi Tu-shuo《新制諸器圖說》of 1627 (Newly Illustrated Manual for the Construction of Machines).

[a] Archeological Research Institute, Kao-Gu-Xue Ji-Chu《考古學基礎》(Foundations of Archeology) (Science Publ. Co., 1958), p. 88.

[b] Song Bo-Yin 宋伯胤 et al., 'Cong Han Hua-Xiang-Shi Tan-Suo Han-Dai Zhi-Ji Gou-Zao'《從漢畫像石探索漢代織機構造》('Investigation of Han Spring Wheel Based on Engraved Stone Picture of Han Dynasty') Wen Wu《文物》(Cultural Relics) 3, 23-30 (1962).

[c] Xia Nai 夏鼐 and Yin Bao-Zhang 殷寶璋, 'Hupei Tonglu-Shan Gu Tong-Kuang'〈湖北銅綠山古銅礦〉, ('Ancient Copper Mine in Tonglu Mountains in Hubei Province') Kao-Gu Xue-Bao《考古學報》(Journal of Archeology) 1, 1-14 (1982).

From the archeological finds and historical records given in Table 1, it can be seen that in ancient China the sliding bearing was first used in the cart and later in the treadle-operated tilt hammer and the water-operated tilt hammer. Hence, the date of the appearance of the cart was probably also the date on which the sliding bearing was born. The sliding bearing on the cart consisted of a wheel boss and a journal fixed on the frame. This should be considered the beginning of the history of lubrication in China. In the development of tribology in ancient China, the sliding bearing on the cart was very important because of its heavy load and high speed.

II. The Appearance of Sliding Bearing Lubrication

Determining the date of the invention of the cart is not difficult because its appearance would have readily drawn public attention. The use of the lubrication, on the other hand, is difficult to observe. Therefore, it is not easy to prove the date of the appearance of lubrication. The earliest record of lubrication is found in the Shi Jing 《詩經》 (The Book of Odes). Shi Jing is the earliest collection of poems from ancient China. The poems collected were those written in the period from the beginning of the Zhou 周 dynasty, to the Chun-Qiu 春秋 (Spring-Autumn) period, from -1100 to -600. There is a poem in the book entitled 'Bei-Feng Quan Shui' 〈邶風·泉水〉 ('The Bei Wind and Spring Water').

> The grease used is sufficient to lubricate the axle
> shaft.
> On the shaft end, we carefully check the pin bolt.
> Quickly, send me home driving the cart.
> Quickly, send me back to my Wei country town.
> Don't let me feel a twinge of conscience.

In addition, there are also records of lubrication in the poem entitled 'He Ren Si' 何人斯 in the 'Xiao Ya' 〈小雅〉 section of the Shi Jing. In this poem, it is stated that only by lubricating the sliding bearing would the cart run quickly.

Nowadays, it is well known that when wood is used as the frictive material, the use of lubrication can affect the coefficient of sliding friction by several orders of magnitude. Since the poems in the Shi Jing were collected and edited from folk songs, it is obvious that by that time lubrication was widely applied and knowledge of it was quite common. In ancient times, it might have taken a very long course to get to this stage, and actually, the date of the appearance of lubrication might be much earlier than recorded in the Shi Jing.

By the time of the Chun-Qiu period, special officers were established for the management of lubrication. As recorded in the Zuo Zhuan 《左傳》 (Master Zuo's Spring-Autumn Annals), the management of lubrication in the palace of the Lu 魯 state and Qi 齊 state during the Chun-Qiu period was the responsibility of the special officers called the Jin-cha 巾車.

III. The Early Lubricant

It is necessary to further analyze what the type of lubricant 'zhi' 脂 (grease) mentioned in the previous poems really was. According to the Shuo Wen Jie Zi 《說文解字》 (Analytical Dictionary of Characters) of +121 by Xu Shen 許慎,

> the word 'zhi' 脂 stands for fats from horned animals, and the word 'gao' 膏 (ointment) stands for fats from non-horned animals.

The common grease at the time of the poem might have actually been 'gao.' This assumption can be explained as follows. First, oxen have been used for pulling carts in ancient China for 4000 years, and for plowing fields for 7000 years. In this sense, the ox was an important source of animal power and could rarely be slaughtered. Second, sheep were widely accepted as food in ancient times, and the saying "lamb and delicious wine" was commonly used. The character for the word "delicious" in Chinese is 'mei' 美 evolved from the two characters 'da' 大 (big) and 'yang' 羊 (sheep). Since the people were fond of eating mutton, sheep fat could have been supplied in abundance.

Moreover, lard was used as a lubricant at a very early date. It is recorded in the Shi Ji 《史記》 (Historical Records) of -90:

> 豨膏棘軸，所以爲滑也。
>
> 'Xi gao' 豨膏 lubricated the sour jujube shaft
> and made the cart move with ease.

The 'xi gao' is lard.[1] All of the records mentioned above indicate that the earliest lubricants were easily obtainable animal fats.

Later, the characters 'zhi' 脂 and 'gao' 膏 took on additional meanings. According to the book Zheng Zi Tong 《正字通》 (Dictionary of Characters) written by Zhang Zi-Lie 張自烈 during the Ming dynasty, 'zhi' refers to solid oily materials while 'gao' refers to melted oily materials. This shows that the state of the substance was a criterion for classification. This point of view is quite similar to our current definition. At present, solid lubricats are called 'zhi' 脂 while liquid lubricants are called 'you' 油 ('oil') rather than 'gao' 膏 . The use of vegetable oil as a lubricant occurred probably late in China.

IV. The Earliest Lubricating Method and the Lubricating Device

Under normal conditions, the bearing should work in a semi-fluid state. However, the melting points of animal fats, especially those of ox and sheep fat, are very high. At normal temperatures, they are in the solid state and would be too hard to be used as lubricants. How then was the lubrication performed? The answer is found in the Shi Ji《史記》. We have a statement that reads:

> The wheel boss and 'guo' 輠 were baked [2]

where 'guo' is a kind of vessel containing the lubricant for the cart. The 'guo' containing lubricant was usually hung on the cart. When lubrication was to be done, the 'guo' and the shaft had to be heated and the hot lubricant in the 'guo' had to be poured on the wheel boss and shaft.

V. The Application of Mineral Oil

Though edible oil (primarily animal grease) was widely used for lubrication in ancient China, there is evidence that China was one of the earliest countries to also use mineral oil for lubrication. The earliest known record is from about +300 in the book Bo Wu Zhi《博物志》(Record of the Investigation of Things) written (c. +270) by Zhang Hua 張華 during the Jin 晉 dynasty. In this book, the following is recorded:

> 酒泉延壽縣南山出泉水，大如莒，注地如溝。
> 取著器中，始黃而後黑，如凝膏。
> 燃極明，與膏無異。
> 膏車及水碓甚佳，彼方人謂之石漆。
>
> The southern mountain of Jiuquan 酒泉 in the
> Yanshou 延壽 county, produced a kind of spring
> with an upwelling as large as a bamboo sprout.
> The gushing spring formed a brook on the ground.

> Collected in a container, its color changed from yellow to black as it congealed into a grease paste. Just as grease fat does, it gave out light when burned. It was an excellent lubricant for carts and water-operated tilt hammers. The natives called it 'shi-qi' 石漆 (literally, 'stone-lacquer').

'Shi-qi' is one of the early names for petroleum. This account is probably the earliest extant record of petroleum.[3] It is of interest to note that, during the Song 宋 dynasty, Shen Gua 沈括 predicted in his monumental work Meng-Qi Bi-Tan《夢溪筆談》(Meng-Qi Essays) that 'petroleum' would be very useful to the world in the future. How right he was!

VI. Wooden Sliding Bearings

Wood was the main material used in machine manufacture in ancient China.[4] The early Chinese developed techniques for selecting the proper wood and for making structural designs to overcome the friction of the sliding bearing. These techniques were of great importance since metal in the early days had not replaced wood as the primary manufacturing material for machines. Even now, in the remote or outlying districts of China, there are some carts in which all parts are made of wood. Without exception, the sliding bearing in ancient China was made of carefully selected wood.

In the quotation from the Shi Ji, "the 'xi gao' lubricated the sour jujube shaft to make the cart move with ease," the "sour jujube" was a kind of high quality hardwood. Actually, quality specifications for the wood of several main parts in the cart were analyzed in the famous Kao Gong Ji《考工記》(The Artificers Record).[5]

In the Ming 明 dynasty, Song Ying-Xing 宋應星, in his book 'Tien-Gong Kai-Wu'《天工開物》(The Exploitation of the Works of

Nature) of 1637, made a comparison of various woods. In Chapter 9, 'The Cart,' it is stated

> In selecting the wood for making a cart, long material should be used for the shaft and short material for the wheel boss. The Chinese scholar-tree, the jujube, the wingceltis and the elm are the fairly good materials. The wingceltis can generate heat if it works continuously for too long. Consequently, it should be used with care. The best way is to use the thick jujube and the thick Chinese scholar-tree.

What is of interest to tribology is the statement that 'the wingceltis can generate heat if it works continuously for too long.' This observation provides a known system for the investigation of heat generation by wooden sliding bearings, a problem which is still of importance in the study of sliding bearings. A possible explanation of why the wingceltis generates more heat than the other woods mentioned probably lies in its xylem which is so hard and dense that it could not absorbe the lubricant well. As a result, the semifluid lubrication would be exhausted if the cart worked continuously for too long.

VII. The Application of Metal Axle Bushings

The use of metal for the axle bushings of the sliding bearing was another important accomplishment in the history of tribology in ancient China. Written records of metal axle bushings are found in the book Wu Zi 《吳子》 (The Book of Master Wu) written by the strategist, Wu Qi 吳起 in the -4th century of the warring states period. In this book, we have

膏鐧有余，則車輕人。

If the 'jian' 鐧 is well oiled, the cart runs lighter than man.

According to the Shuo Wen Jie Zi, the 'jian' 鐧 is the iron outside the axle and the 'gong' 釭 is the iron inside the wheel. A similar description is given in the Shi Ming 《釋名》 (Exposition of Terms) written by Liu Xi 劉熙 during the Han dynasty: "the 'jian' is placed between the 'gong' and the axle to prevent them from producing friction with each other."

In April 1974, in Mingchi 澠池 county, Henan 河南 province, a storage of ancient iron products was discovered, including hundreds of iron axle bushings.[6] According to initial studies, the date for the production of these artifacts is earlier than the Northern Wei 魏, about the 5th century. The interior structure of these axle bushings is in the shape of a circle with two or three tenons. The exterior structure is in the shape of a semicircle (with one tenon) or a hexagon (see Figure 1). The hexagonal shape of the tenon acted to prevent circular

Figure 1. Ancient 'gong' 釭 (iron axle bushings) unearthed in Mingchi 澠池 county, Henan 河南 province in April 1974 (Ref. 6).

motion by the bushings. These axle bushings were placed in the wheel bosses (called "gong" in the reference).

A Han dynasty kiln for baking molds was discovered in Wen 温 county, Henan province, in May 1974.[7] There were hundreds of pottery mold sets for axle bushings. The structure was basically the same as the above mentioned axle bushings (see Figure 2).

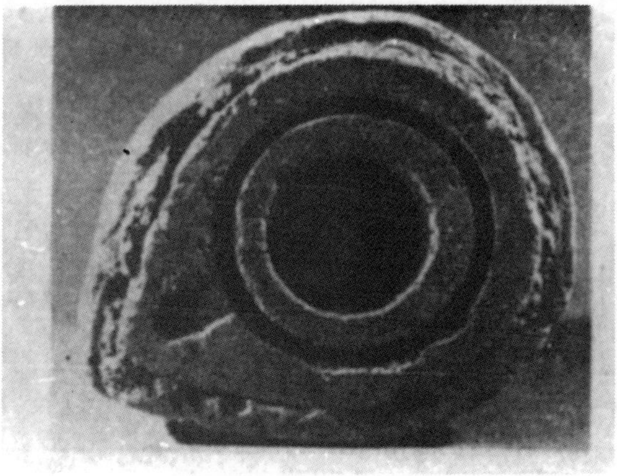

Figure 2. A Han dynasty axle bushing mold unearthed in Wen 温 county, Henan 河南 province, in May 1974 (Ref. 7).

Thus it can be said that the metal axle bushing appeared approximately 2400 years ago in China. The axle bushing was made of cast iron, low carbon steel or copper. The structure could be in two layers: the interior layer called the 'jian' and the exterior layer called the 'gong' (see Figure 3). The structure also could be in one layer. During the Han dynasty, the 'jian' was used quite widely and was produced in large batches necessitating the use of standardized measurements.

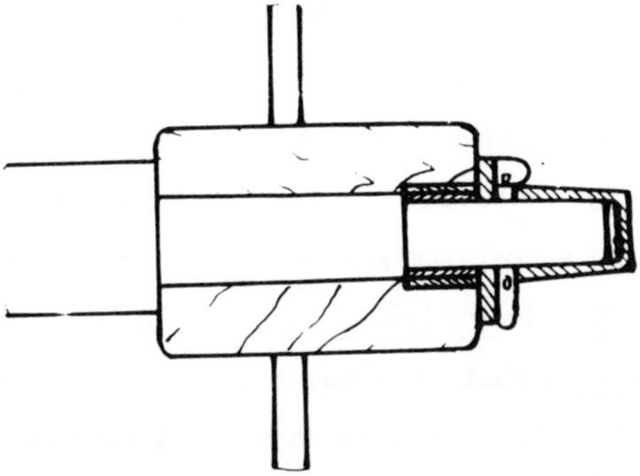

Figure 3. An imaginary sketch of the ancient 'gong' 釭 and 'jian' 鐧 .

VIII. Sliding Bearing Structure Research in Ancient Times

Through improved structural designs, the early Chinese were able to reduce the friction and wear of sliding bearings. By the time of the Warring States, it was well understood that the ratio of the length and the diameter of the sliding bearing must be correct, otherwise, the sliding bearing could not work well. We have from the Kao Gong Ji《考工記》the description:

> When the diameter of the wheel boss is small and the length of the sliding bearing is large, the resistance is greater. Conversely, when the diameter of the wheel boss is large and the length is small, the sliding bearing is not stable enough. When the length of the wheel boss is small, the resistance of the sliding bearing will be small; on other hand, when the length of the wheel boss is large, it will be safer (i.e. stronger).

Much of the understanding of the structural designs of sliding bearings undoubtedly derived from early work on chariots. The impressive and dignified manner and the rapid speed of the chariot were often praised by early writers.[8] In the work of Zhu Xi 朱熹 of the Song dynasty, we find a comparison between the chariot and the cart. He points out in his Shi-Jing Ji-Zhu《詩經集注》(Annotated Collection of Poems and Odes) that the chariot, in the size of its whole body, was obviously smaller than the cart (a kind of large-scale transport vehicle), but the length of the wheel boss of the chariot — the dimension of the sliding bearing in the axial direction — was longer than that of the cart (see Figure 4). The length of the wheel boss of the cart was one and a half 'chi' 尺 (Chinese feet) and that of the chariot was three point two chi. This information was also cited in the book Tian-Gung Kai-Wu《天工開物》(The Exploitation of the Works of Nature) of 1637. As mentioned above, in order to meet the requirements of going on expeditions and galloping, the chariot had to be built even sturdier. Therefore, the load on the sliding bearing on the chariot was greater as its speed increased. Thus, it should have a much higher load-carrying capacity. It seems that the ancients

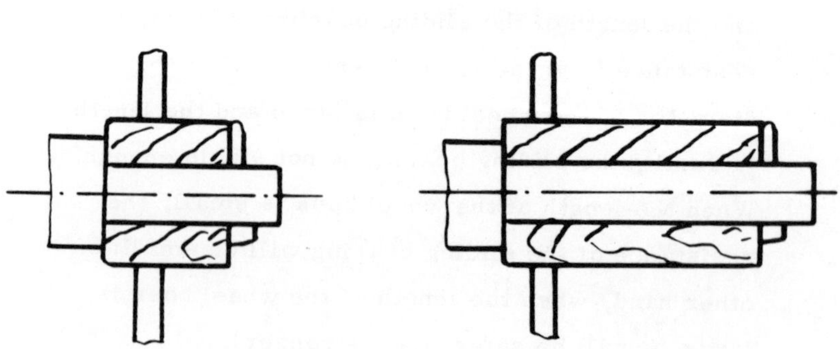

(a) (b)

Figure 4. Comparison of the lengths of the sliding bearings of the cart (a) and chariot (b).

understood that the only way to meet the requirements of the chariot was to increase the length of the wheel boss and not increase the bearing diameter. This conclusion is in accordance with that obtained based on the pressure ratio and the friction power ratio analysis of semifluid lubrication of the slide bearings. When the bearing diameter is increased, the pressure ratio will be decreased, but the friction power ratio cannot be reduced and the load-carrying capacity of the slide bearing may not be increased. Only when the length of the wheel boss is increased properly can the pressure ratio and the friction power ratio be reduced at the same time. At high sliding velocity, the effect is significant since the friction power ratio increases with velocity.

IX. Replacing Sliding Friction with Rolling Friction

In order to reduce friction and wear, it is effective to replace sliding friction by rolling friction. The use of transversely placed wooden rollers to transport huge stones began very early in antiquity. Later, iron rollers were used. A well-known example is the transportation of huge logs from Jing Xi 江西 to Le Yang 洛陽 for constructing the palace during the Sui 隨 dynasty around 600. An account of the event is found in the Zhen-Guan Zhang-Yao《貞觀政要》(Records of Zhen-Guan Reign) by Wu Jing 吳竟 of the Tang 唐 dynasty. It is stated that only by putting iron rollers under the wood can two thousand people move it by pulling. If wooden rollers were used, the friction would have resulted in fire.

A significant step in the use of rolling devices to reduce friction is the invention of roller-bearings. The earliest extant record of roller-bearings is found in the Tian-Wen Zhi〈天文志〉of the Yuan-Shi《元史》(History of the Yuan dynasty) of 1370 in which the astronomical instruments of Guo Shou-Jing 郭守敬 are described. It is in the instrument 'jian yi' 簡儀 ('equatorial torquetum' or literally 'simplified instrument') constructed in 1270 that four roller-bearings were installed to allow the easy rotation of the equatorial ring against the fixed diurnal ring, graduated in the 12 double-hours and the 100 quarters with 36 sub-divisions.

Acknowledgments. In the course of this research, I have received the advice of Professor Chen Lin-Chin 鄭林慶 and have been helped by Professor Wang Ze-Shen 王澤申, Wang Bin-Quan 王秉銓 and Zang Ji-Chang 臧吉昌. I hereby express my thanks.

References

1. This interpretation of the term 'xi gao' is the same as that given in the Shuo Wan Jie Zi《說文解字》(Analytical Dictionary of Characters) of +121.
2. See the section 'Menz Zi Xun-Qing Lie-Chuan'〈孟子·荀卿烈傳〉('The Biographies of Master Menz and Master Xun') in the Shi Ji《史記》(Historical Records).
3. The place Jiuguan 酒泉 where the earliest extant record of petroleum was made is the natural oil well of Yuman 玉門 in Gansu 甘肅 province.
4. Copper began to appear about 4000 years ago as a manufacturing material in China. Iron began approximately 3000 years ago.
5. The Kao Gong Ji《考工記》(The Artificers Record), incorporated in the Zhou Li《周禮》(Record of the Chou Rites) of -140, was written before the Qin 秦 dynasty.
6. 'Mingchi Xian Fa-Xian de Gu-Dai-Yao Cang Tie-Qi'〈澠池縣發現的古代窰藏鐵器〉('Iron Axle Bushings Discovered in an Ancient Storage Site at Mingchi County') by Mingchi-Xian Wen-Hua Kuan 澠池縣文化館 ('Culture Center of Mingchi County') in Wen Wu《文物》(Cultural Relics) 8, 44-60 (1976).

7. 'Henan Wen Xian Han-Dai Hong-Fan-Yao Fa-Jue Bao Gao' 〈河南温縣漢代烘範窰發掘報告〉('Report on the Discovery of a Han Kiln in Wen County of Henan Province') by Henan 河南 Museum in *Wen Wu* 《文物》(*Artifacts and Civilization*) *9*, 66-75 (1976).

8. See for example, the poem entitled 'Qin-Feng Xiaorong' 〈秦風·小戎〉 in the *Shi Jing*《詩經》(*The Book of Odes*).

GUNPOWDER/ROCKET TECHNOLOGY IN ANCIENT CHINA AND ITS TRANSFERENCE TO THE OUTER WORLD*

Fang-Toh Sun 孫方鐸

I. Introduction

It is generally recognized that the Chinese have a long history of gunpowder development which ultimately led to the invention of the rocket. But there are also some doubts about such a belief. In this paper the historical development of gunpowder/rocket technology in ancient China from the -2nd century to the +17th century will be closely reviewed and the findings and opinions of the many contemporary researchers on this subject duly considered. With such a survey as a basis, the following problems will be discussed from a comparative perspective:

1) Can the development of gundpowder/rocket technology in ancient China be regarded as independent?
2) If it was independent, was it unique, or have there been parallel developments in the other parts of the world?
3) How did such early technology in China transfer to other countries?

The purpose of this study is fact-finding. To help to obtain a correct perspective, a chronological table is constructed to show the datable events concerning such developments in ancient China as well as in the outer world. Owing to the extreme difficulties in gathering the materials and even more in interpreting them due to the unavoidable language barriers, the present study is by no means comprehensive, and the chronicle table so constructed is very preliminary.

---

*Presented at the 17th International Congress of History of Science, University of California, Berkeley, August 1985.

Nevertheless it is hoped that the historical facts found so far may provide at least some ground for further study of this difficult subject.

II. The Origin and Development of Gunpowder/Rocket Technology in Ancient China

A. The Origin of Gunpowder

As we know, the three principal ingredients of gunpowder are sulphur, saltpeter and charcoal, hence the natural occurrence of all three of these substances is a prerequisite for the origin of gunpowder in a particular country. China is lucky in this regard, as it is rich in both sulphur and saltpeter, while charcoal is a common domestic material in China. The records show that natural sulphur is found on the eastern seacoast in the mountainous regions of Tai 太 and Mu-yang 牧羊, and in the region west of the Yellow River, as well as in the southern provinces, such as Hunam, Kwangtung, etc., while natural saltpeter is found in Shangtung, Shanshi, and Szechuan.[1] In fact, the name sulphur ('liu huang' 硫黄) appears early in Prince Liu An's 劉安 book, Huai-Nan Tzu《淮南子》(The Book of the Prince Huai Nan)[2] of about -120; and it is also said that the Prince purified the 'ch'iu shih' 秋石 (literally, autumn mineral),[3] in his search for elixers, according to a later book, the Chou-I Ts'an T'ung Ch'i《周易叅同契》written by the pioneer alchemist, Wei Po-Yang 魏伯陽 in about +142; and the so-called 'ch'iu shih' is believed to be 'hsiao shih' 硝石, saltpeter.[4] Thus both sulphur and saltpeter were known to the Chinese at least by Liu An's time.

With the abundance of sulphur and saltpeter in China, the ancient Chinese people became acquainted with their properties and mastered the technique of their purification or distillation through medical studied and alchemical experiments. In the Shen-Nung Pen

Ts'ao Ching 《神農本草經》(Classical Pharmacopoeia of the Heavenly Husbandman), one of the earliest medical books written in the second century, saltpeter is listed among the 120 kinds of medicines in the superior class, while sulphur is listed among the 120 kinds in the medium class.[5] In the Chin-Kuei Yao-Lüeh 《金匱要畧》(The Essentials in the Golden Chest) a classical work of the famous physician Chang Chung-Ching 張仲景 of the Later Han dynasty in the second century, saltpeter appears in several medical recipes for the curing of some particular diseases, such as jaundice, for example.[6] In a later book, the Pei Chi Ch'ien Chin Fang 《備急千金方》(A Thousand Golden Recipes for Emergency Uses), written by the famous Taoist physician Sun Ssu-Mo 孫思邈 of the T'ang dynasty in the 7th century, both saltpeter and sulphur appear in a number of prescriptions. As to alchemical applications, sulphur and saltpeter were the most important ingredients in the making of elixers. They appear in various formulas in the Taoist's books on alchemy, from the earliest book, the Chou I Ts'an T'ung Ch'i of +142 by Wei Po-yang to Ko Hung's 葛洪 Pao-P'u Tzu 《抱朴子》(Book of the Preservation-of-Solidarity Master) of 320, Sun Ssu-Mo's 孫思邈 Tan-Ching Yao-Chüeh[7] 《丹經要訣》(Essentials of the Elixir for Oral Transmission) of 640, and many others. As mentioned before, both these two ingredients had been associated with Prince Liu An in his research on elixirs as early as about -120. One particular piece of evidence to show the popularity of saltpeter preparations is the fact that the smelting of saltpeter during the period from the day of the summer solstice until the beginning of autumn was prohibited by government law on the ground of safety, as recorded in Fan Yeh's 范曄 Hou Han Shu 《後漢書》(History of the Later Han Dynasty).[8] This was about the first to second century B.C.

B. From Alchemical Mixture to Proto-gunpowder

With natural sulphur and saltpeter discovered, their characteristic properties studied, and the techniques of their handling and treatment mastered, the stage for the invention of gunpowder was set at the latest by the 4th century in China. However, no true gunpowder appeared in China until several hundred years later. The first record of a gunpowder-like mixture is found in the book Chu Chia Shen P'in Tan-Fa《諸家神品丹法》(Methods of Various Schools for Magical Elixir Preparation) which is a collection of various alchemical processes. It was compiled by Meng Yao-Fu 孟要甫 of the Northern Sung dynasty, but the various processes recorded could have originated much earlier. In a particular process, called 'fu huo liu huang fa' 伏火硫黃法 (literally, 'a method for subduing sulphur by fire') described in this book, the mixture consists of equal amounts of sulphur and saltpeter, to which are added some heated pods of the soap-bean tree, serving as the source of carbon. The purpose of this process was not to deflagrate, but, as the name suggests, to subdue sulphur. But all three of the principal ingredients of gunpowder were actually there. The reason that it did not deflagrate is probably due to the low percentage of carbon, and because the process was carefully controlled. As indicated in the book this process is attributed to a man named Sun 孫, with the title of Chen-Jen 眞人, literally, 'the supernatural man,' a title reserved for distinguished Taoists of high achievements. The true name of this man and his lifetime are still under dispute. Some take him to be the famous Taoist, Sun Ssu-Mo, but others doubt it.[9] In particular, Chen, judging from the fact that several geographic names which appear in the description are not those in current use in Sun's time but about one hundred years after, concludes that this book must have come out a hundred years later, and that the author was not Sun Ssu-Mo, but someone else.[10] Despite the uncertainty regarding the identity of the true contributor of this process, it is reasonable to conclude that such a gunpowder-like mixture appeared in China in about the 8th century. A similar process is found in another Taoist's book, the Ch'ien Kung Chia Keng Chih Pao Chi Ch'eng《鉛汞甲庚至寶集成》(The Complete Compendium on the Perfected Treasure of Lead, Mercury, Wood and Metal). As described in this book, the process is called 'fu huo fan fa' 伏火礬法 ('a method for subduing alum (or vitriol) by fire') and the mixture also consists of equal amounts of sulphur and saltpeter, as in the previous process, but added to aristolchia as the carbon source, instead of the pods of the soapbean tree. These three ingredients, after being pulverized and thoroughly mixed, are put into a pot, buried underground, and then heated. In the end the sulphur is subdued, ready for its use as the agent to further subdue the alum or vitriol. As given in the book, this process is attributed to a Taoist, named Chin-Hua Tzu 金華子 in the year 808. Just like that by Sun Chen-Jen 孫眞人, this process is not aimed at deflagration, but it does comprise the three principal ingredients of gunpowder. Hence, among the numerous alchemical mixtures found in Tao Tsang《道藏》(The Taoists' Patrology) these two particular mixtures appear to be the prototypes of gunpowder.

As to the first appearance of a deflagrating gunpowder-like mixture, a record is found in another Taoist's book, the Chen Yüan Miao Tao Yao Lüeh《眞元妙道要畧》(The Classified Essentials of the Mysterious Tao of the True Origin of Things). In this book it is said that the heating of sulphur together with realgar, saltpeter, and honey had once produced flames so serious that their hands and faces were scorched, and their houses burned to ashes. Here both sulphur and saltpeter were present, with carbon provided by the honey, and some extra sulphur from the realgar. This book is attributed to a

Taoist named Cheng Ssu-Yüan 鄭思遠, who had also the title Chen-Jen. Just who this man was and when his lifetime was is a problem which needs investigation. He could be the man from whom Ko Hung 葛洪 learned his alchemy as mentioned in the Pao P'u Tzu, as some people believe. If so, then he was a man of the Three-Kingdom period, and this book must have been written in the early 3rd century. However, according to Weng's recent study,[11] there were two Taoists in the T'ang dynasty having this same name, and besides, in the same book several references are made to people of the T'ang dynasty, and even some of the Later Chin dynasty in the Wu-Tai 五代 period. So the time when this book was written must be shifted down to about seven hundred years later. Regardless of who the true writer of this book was, it is reasonable to say that such a deflagrating gunpowder-like mixture first appeared in the middle of the 10th century, or earlier. Of course, the emergence of such a deflagrating mixture was accidental, not intentional, since the Taoists' aim was to prolong man's life, not to kill him, and such a deflagrating property was warned as dangerous, and to be safeguarded.

C. The Beginning of the Gunpowder Age

At just what time such a life-prolonging mixture in the alchemist's hands turned to being the destructive gunpowder for military use, we do not know. But not long after the appearance of such proto-gunpowders from the late years of the T'ang dynasty to the Wu-Tai period, three recipes for military gunpowder appeared in the official publication of the Northern Sung dynasty, the book Wu Ching Tsung Yao《武經總要》(Collection of the Most Important Military Techniques) compiled by Tseng Kung-Liang 曾公亮 and Ting Tu 丁度 by imperial order, and completed in 1044. The official name adopted there for such gunpowder mixture was 'huo yao' 火藥 (literally, 'the fire drug'). Thus the transition from alchemy to armament seems to have happened sometime between the late 9th and the early 11th century. A comparison of the compositions of these early Chinese gunpowders with those of other countries in later years, and that of the modern black powder is given in one of the author's earlier papers,[12] and for the convenience of the reader, it is reproduced below with some slight modifications (Table I). In the table, only the relative amounts of the three principal ingredients, sulphur, saltpeter, and charcoal are given, while the many additives are neglected. Detailed descriptions and complete recipes including all of these additives are found in the Wu Ching Tsung Yao, and the English translations of two of them are given by Wang Ling 王鈴.[13]

Together with these three recipes for military gunpowder, various forms of gunpowder-powered firearms are also found in

Table I. Compositions of the Early Chinese, Arabic and European Gunpowders (in percentage by weight)

	Chinese (1044)[a]			Arabic[b]	European		Modern[e]
	A	B	C	Hassam er-Rammah (1280)	Bacon[c] (1247)	Magnus[d] (1275)	Black-powder[e]
Saltpeter	61.5	60.0	52.0	69.5	41.2	66.7	61.6
Sulphur	30.8	30.0	27.3	14.8	29.4	11.1	15.4
Charcoal	7.7	10.0	20.7	15.7	29.4	22.2	23.0

[a] Tseng Kung-Liang and Ting Tu, Wu Ching Tsung Yao 1044, Vol. 11, p. 28; Vol. 12, pp. 58, 65. A: barbed fire ball, B: smoke ball, C: 'huo p'ao.'

[b] Willy Ley, Rockets, Missiles and Men in Space, 1968, p. 53.

[c] Ibid., pp. 53-54.

[d] W. R. Maxwell, 'The Early History of Rockets,' (J. British Interplanetary Society 35, 176 (1982).

[e] J. Humphries, Rockets and Guided Missiles (Macmillan, 1956), p.27.

Wu Ching Tsung Yao, such as the 'yen ch'iu' 烟毬 ('smoke ball'), toxic or non-toxic; the 'chi li huo ch'iu' 蒺藜火毬 ('barbed fire ball'), the 'p'i p'i huo ch'iu' 霹靂火毬 ('thundering fire ball'), the 'huo p'ao' 火砲 ('fire catapult'), and many others. Such a great variety of new firearms shows how gunpowder had been widely adapted to military uses by the early years of the Northern Sung dynasty at the beginning of the 11th century, and Chinese armament actually entered its gunpowder age then. Historical records show that these new forms of gunpowder firearms actually went into action during the wars among the Sung, the Chin Tartars, and the Mongols throughout the Sung dynasty. However, owing to the limited scope of the present paper, such gunpowder firearms cannot be fully discussed, except a few of the datable events concerning them are given in the appended chronicle. In the following we will concentrate on the development of rocketry in ancient China.

D. From the Incendiary Arrow to the Gunpowder Rocket

The Chinese name for rocket is 'huo chien' 火箭 (literally, 'the fire arrow'). The use of the 'huo chien' as a war weapon can be traced all the way back to the time of the Three Kingdoms in the 3rd century. However, these early Chinese rockets, attached with inflammable materials, and shot by bows or crossbows, should not be confused with modern Western rockets, which operate on the reaction principle. A careful study of the development of 'hui chien' from the incendiary arrow to the true solid-propellant rocket is given in the author's previous work,[14] and the findings there may be summarized as follows:

1. The early 'huo chien' in China from the time of the Three Kingdoms to the early years of the Northern Sung dynasty (3rd to 10th century) was an incendiary weapon, not a rocket in the modern sense.

2. Such early 'huo chien' became gunpowder-charged from the beginning of the gunpowder age in the early 11th century, but the gunpowder applied to the conventional arrow served only to increase its incendiary power, not for propulsion. Hence, it was still not a rocket.
3. The first rocket-like weapon to appear in China was the 'fei-huo ch'iang' 飛火槍 ('the flying fire lance'), employed by the Chin tarters against the Mongols in the battle of K'aifeng-fu 開封府 (Pienking 汴京)[15] in 1232. This was one of the two brand-new weapons the Mongols dreaded most.[16] It could be regarded as a primitive solid propellant rocket on the following grounds:
   (a) The gunpowder-like mixture was stored in a container attached to the lance instead of being directly applied to the arrowhead as in the gunpowder-charged 'huo chien,' and the flames produced upon ignition erupted from an opening in the container forming a hot jet.
   (b) The lance was launched without a bow or crossbow, and range was gained with the help of the recoil of the erupting gas.
4. After the battle of K'aifeng-fu, not much rocket activity was found during the Yuan dynasty (1277-1368) established by the Mongols in China after the Chin and Sung were successively overthrown. This was probably due to the fact that the Mongols favored the trebuchet and the 'huo p'ao' ('fire catapult') which were of higher destructive power than the rockets. But gunpowder rockets had been reactivated at the beginning of the Ming dynasty after Emperor T'ai Tsu 太祖 drove the Mongols out of China.

5. In the Ming dynasty (1368-1644) a great variety of rocket weapons were developed under the old general name 'chien' with various modifiers according to the design, and good progress was made in many aspects, notably:

   (a) The separation of the gunpowder into two distinct portions; the 'ts'ui huo' 催火 ('pushing fire'), stored in a container near the rear of the arrow's shaft, responsible mainly for propulsion, and the 'shao huo' 燒火 ('incendiary drug'), stored in a container close to the arrowhead, responsible mainly for the production of fire. These two portions of gunpowder were different in composition since their functions were different, the former being higher in saltpeter and the latter, higher in sulphur.

   (b) The projectile range was advanced from one hundred paces[17] with an accompanying increase in the arrow's size and weight, and stronger arrowheads of specially designed shapes were introduced.

   (c) The development from the single rocket to multiple rockets, ranging from two or three rockets in one unit, as in the 'shen-chi chien' 神機箭 ('magic arrow'), or up to a hundred in one package, as in the 'pai-hu ch'i-pen chien' 百虎齊奔箭 ('hundred tigers racing abreast'). The destructive power was greatly increased.

   (d) The introduction of many diversified designs for rocket weapons, including winged rockets (a miniature missile), and the rotating rocket (a defensive weapon) called 'wan jen ti' 萬人敵 ('a match for ten thousand men').

Details of such developments, together with a list of 19 rocket weapons of the Ming dynasty are given by Sun.[18]

6. The active use of gunpowder rockets in the Ming dynasty is evidenced by the following historical facts:

   (a) A particular kind of gunpowder rocket, the 'shen chi chien' was listed among the standard armaments for national defense in the early years of the Ming dynasty under its first Emperor T'ai Tsu.[19]

   (b) The actual deployment of gunpowder rockets, in one form or another, is found in historical records concerning the campaigns of Chinan 濟南 (1400),[20] Luchuan 麓川 (1464),[21] and Ta-t'eng Hsia 大藤峽 (1465),[22] for example.

   (c) The training of soldiers for the testing and operating of gunpowder rockets was conducted by the famous General Ch'i Chi-Kuang 戚繼光, and directions for making such rockets are given in his book, the Chi Hsiao Hsin Shu 《紀效新書》 (A New Treatise on Military and Naval Efficiency).

From the materials presented above, we see clearly that the Chinese term 'huo chien' possessed different meanings in different epochs as follows:

10th century and earlier: ordinary incendiary arrows
11th to 13th century: gunpowder incendiary arrows
13th century and after: gunpowder rockets

The most confusing period is the time around 1232, when the gunpowder rocket made its first appearance. While the Chin Tartars regarded it as a kind of lance, the Mongols called it 'p'en huo t'ung chien' 噴火筒箭 ('the arrow with a fire-erupting tube'). However, soon after this short period, the general name 'chien' 箭 was restored during the early years of the Ming dynasty, as mentioned in the preceding paragraph.

After the middle of the 16th century, as the barrel guns and cannons introduced from the West gained more favor from the government, the development of gunpowder rockets in the Ming dynasty seems to have stalled. In fact, a few of the Western weapons such as the famous 'hui-hui p'ao' 回回砲 had been brought to China by the Mongols in the 13th century, and in the 16th century the muskets and bird-beak guns from the Portuguese were also found in China. The ever increasing Western influence on Chinese firearms in the Ming dynasty can be seen from Chiao Hsü's 焦勗 book The Tse K'o Lu 《則克錄》 (Elements of Fire Warfare) published in 1643. It was actually the work of Schall von Bell, one of the foreign priests who came to serve in the Chinese court, but was written by Chiao, based on von Bell's oral presentation. Though the use of gunpowder rockets was occasionally reported thence, evidently, it was no longer a popular weapon in current use.

E. The Development of Pyrotechnics

Ever since the beginning of the gunpowder age in China, its applications have been going in two directions: firearms for the military, and fireworks for amusement. With the gunpowder firearms presented above, a few words about the fireworks may now be said:

1. The Chinese name for firework is 'yen huo' 烟火 which means smoke and fire. Like the name 'huo chien,' 'yen huo' has also a long history in China. It is supposed to have prevailed in China as early as the Sui dynasty in the 6th century. But according to Chao 趙[23] such early 'yen huo' had nothing to do with gunpowder.
2. Advanced fireworks using a gunpowder-like mixture prevailed in the Southern Sung dynasty in the early years of

the 12th century, and spectacular fireworks were often displayed for the entertainment of the emperor.

3. Of the various fireworks in the Southern Sung dynasty, the one called 'the ground rat' deserves our particular attention, as it shows clearly the reaction principle by running around under the recoil of the flames produced after ignition. There is a story[24] which says that in a royal party at the palace the wild running of a ground rat frightened the Dowager-Empress by suddenly going underneath her throne. The time for this unhappy incident was in the early years of the reign of Emperor Li Tsung 理宗 of the Southern Sung dynasty around 1226, which was a few years earlier than the battle of K'aifeng-fu. So Needham believes[25] that such ground rats were the first rockets in China. In fact, not only the ground rat, but also many other fireworks had been operating on the reaction principle by that time.

With the gunpowder rocket gradually dying out in China after the Ming dynasty, gunpowder fireworks still enjoy their popularity in all parts of China today. However, a comprehensive presentation of gunpowder pyrotechnics is beyond the scope of the present paper.

F. Conclusions

In summarizing the materials presented above, we see that:

1. From the early recognition of sulphur and saltpeter through alchemical and medical studies in the -2nd century until the formal beginning of the gunpowder age in China in the +11th century, the technical progress leading to the invention of gunpowder went on for over a thousand years, and no record has been found showing outside influence during these long years of development.

2. A similar statement may be made on the development of rocketry, except that the military rocket in China made its first appearance in 1232, whereas the festival rocket, like the ground rat, appeared a few years earlier.
3. Significant Western influence on Chinese firearms was not found until the Ming dynasty in the 16th century, but by that time the gunpowder age in China had already begun well over 500 years earlier, and rocketry had been in progress for over 300 years.

Thus, in conclusion, the development of gunpowder/rocket technology from the early years of ancient China up to the middle of the 16th century can be regarded as independent of developments in other civilizations.

For the sequence of events concerning the development of gunpowder and rocket technology in China as well as in the other parts of the world, see the chronicle table in the appendix.

III. <u>On the Westward Transference</u>

Regarding gunpowder/rocket technology in Europe, it is generally acknowledged that the first written European account of such technology was given by Roger Bacon in his <u>Epistola</u> around 1247[26]; and the first European war rocket in action appeared in the battle of the Iberian Pennisula in 1249.[27] However, by this time such technology was already in wide application in China and nowhere else, as shown by the known historical records (see the appended chronicle table). Thus, the fact that it was transferred to Europe from China in the 13th century or earlier is unquestionable and the most probable routes for such transference were through the Moslem countries in Central and Western Asia, and through India, both situated between Europe and China, and both having highly developed civilizations at that time.

A. The Role of the Moslem Countries

While gunpowder firearms and fireworks were highly developed in China during the 10th to the 13th centuries, the various countries of Central and Western Asia, comprising mainly the Arabs and Persia, were united within the great Moslem Empire, which included even Egypt in Africa and Spain in Europe. So far as science and technology are concerned, the Moslem contributions in mathematics, astronomy, alchemy, and medicine are well-known. In particular, gunpowder/rocket development in these countries in the medieval ages has been extensively studied by Laufer, Ley, Needham, Feng, and many others. The following are some of their findings which have a direct bearing on the possible transference of such technology:

1. The Arabs obtained saltpeter from China in the 13th century, and hence called it 'snow from China.'[28] This same name also appears in an Arabic book written by Ibn Albaither in 1240.[29]
2. The rocket was called 'the Chinese arrow' in an Arabic manuscript, written by Hassan er-Rammah in 1280, in which recipes for making gunpowder and directions for making rockets were given.[30]
3. Many Moslem firearms resembled those used earlier by the Chinese, and hence were possibly developed from them. For example, the Arabic weapon 'mafaa' which appeared in the late 13th century was possibly the descendant of the Chinese fire lance or fire barrel used in battles among the Chinese, Chin Tartars, and Mongols, a hundred years earlier.[31] Many such examples are found in Feng's article.[32]

All of these findings show a strong Chinese influence on firearms in the Moslem countries. This should be of no surprise if we look back at the long history of contact between China and the countries in the vast area of central Asia lying to the west of China proper, accordingly named 'hsi yü' 西域 ('the West Region') in Chinese history. Disregarding the legends about very early contacts, the datable events can at least be traced back to the time of the reign of Emperor Wu Ti 武帝 of the Earlier Han dynasty, when Chang Ch'ien 張騫 led the first expeditions to that region during the years -126 to -115. The result was that 36 states in the area were subdued, including Parthia and Sodiana,[33] and the so-called 'old silk road' began to take shape. Subsequent expeditions led by Pan Ch'ao 班超 took place from +73 to +105 in the Later Han dynasty.[34] During his 30 years in the West Region, more than 50 states were subdued, and one of his envoys even reached the Caspian Sea. The old silk road was thus further developed. While such overland communication continued, maritime routes were also under development. It is said that Chinese ships reached the mouth of the Euphrates in Iraq by the 5th century, and Chinese merchandise appeared at the annual fair of Batanea on the Euphrates around 360. From the 8th century onward, Arab ships came frequently to Chinese sea ports such as Canton 廣州, Hengchow 杭州, and Chuangchow 泉州. Such sea routes played an important role in the connection between the Southern Sung dynasty and the Western world in the 11th to the 13th century, when Northern China was occupied first by the Liaos and Chin Tartars, and then by the Mongols. A full description of such East-West communications is found in Needham.[35]

From such long years of contact, the westward transference of Chinese technology to Europe through the countries in Central and Western Asia (most had become Moslem by the 6th century) is to be expected. So far as gunpowder/rocket technology is concerned, the transference occurred most probably during the period from the 11th to the 13th century when such technology matured in China. The possible channels for such transference were: 1) commercial, 2) missionary, 3) diplomatic, and 4) military. With the numerous accounts, Chinese and Western, on communication through these four channels, a concise review is impossible in the present limited space. In the following we will review briefly only on the role played by the Mongols in such transference, since the 13th century was the time the Mongolian impact was felt all over the East and West.

B. The Role of the Mongols

As we know, the warlike Mongols favored the trebuchet which was a stone-throwing device, not gunpowder-powered, better than other weapons they had. Hence, some scholars, like Feng and Needham, think the Mongols could not have been responsible for the transference of gunpowder/rocket technology.[36] However, the following facts cannot be overlooked:

1. The Mongols did employ gunpowder firearms during their three massive westward invasions in the 13th century. A few historical accounts may be cited as follows:
   (a) During Genghis Khan's invasion of Khwarizm, General Kuo Pao-Yü 郭寶玉 fire 'huo chien' in his pursuit of the defeated King Ala ad-din Muhammed, when he flew across the River Amu Darya in 1220.[37] Since this was the time after the dawning of the gunpowder age, but still in the pre-K'aifeng-fu stage, such 'huo chien' could not have been true rockets, but it must have been the gunpowder fire arrows like those widely used by the Hans and Chin Tartars since the 11th century.
   (b) During the campaign of Khrosan in 1221, Prince Tulai, the youngest son of Genghis Khan, deployed 700 'huo kuan' in his attack on the City Nishapur.[38] So far as

the name 'huo kuan' 火罐 (literally, 'fire pot') is concerned, it appears in the book, Wu Ching Tsung Yao as a device used in conjunction with the so-called 'meng huo yu chi' 猛火油機 ('fierce oil machine') which employs mainly petroleum, but is assisted by gunpowder. Whether those employed by Tulai were identical to what was described in the Wu Ching Tsung Yao is not certain, but presumably they were a similar sort of device.

(c) During the battle of Liegnitz in 1241, it was reported that there appeared in the Mongolian army a mysterious monster which spurted out stinky smoke that caused the defeat of the allied army of Poland and Germany.[39] This was a sort of poisonous smoke ball, a Chinese gunpowder firearm also found in the Wu Ching Tsung Yao, according to Feng.[40]

(d) During the siege of Baghdad in 1258, the Mongolian army under Hulagu Khan used a weapon, called the 't'ien p'ing' 鉄瓶 (literally, 'the iron bottle') against the garrison force of the Caliph. It is believed to have been a sort of grenade that first appeared in the battle of K'aifeng-fu 26 years before under the name 'heaven-shaking thunder,' according to some French gunpowder specialists.[41]

These historical accounts show that, even though nothing is said about the use of rockets by the Mongols, at least gunpowder firearms were employed during their military actions in Central and Western Asia and in Eastern Europe in the period 1220 to 1258. This should be expected since this was a time when gunpowder firearms were well developed in China and widely applied during the battles among the

Hans of the Southern Sung dynasty, the Chin Tartars, and the Mongols.

2. The mere employment of gunpowder firearms in a region may not mean the transference of such technology to the people in that region, except in raising their attention by the impressive manifestations of the terrible power of such new weapons. However, it was the usual practice of the Mongols that, in their relentless slaughter of their enemies - soldiers and civilians alike - after the capture of each city, the technicians were spared and organized to become a supporting force for their military operations. Thus it is no wonder that new weapons such as the gunpowder firearms which the Mongols encountered on their eastern front against the Chin Tartars and Hans, were quickly employed on their western front against the Moslems and the Europeans. Besides, with the large concentration of military technicians of many different nationalities including Arabs, Persians, and Europeans, in addition to the Chinese in the Mongolian army, there must have been ample opportunities for technical interchanges.

3. Under the great Mongol Empire formed from the 13th to the 14th century comprising the whole of Asia and a part of Europe, communication between the East and West reached a heretofore undreamed of new stage, and it was reported that travel along the overland road from Eastern Europe to China was perfectly safe by day and by night.[42] Such easy travel certainly enhanced the contacts between China and the West through all channels, commercial, missionary, and diplomatic, as well as military, and thus promoted the

transference of all kinds of technology, which undoubtedly included gunpowder and rocketry.

C. The Role of India

The popularity of festival rockets in India, and the high level of activity in war rockets found in some periods in Indian history is well-known. With the early development of high divilization in the Indus valley, there is some speculation that India was the birthplace of rockets rather than China. Some scholars believe that the local tribe, known as the Oxydracae, which repulsed the army of Alexander the Great near the River Beas in 325 used rocket-like weapons. Some go even further by regarding the words 'agni astra' found in ancient Vedic hymns composed a couple of thousand years B.C., and the word 'vana' or 'bana' found in the great Indian epic poem, Ramayana, composed around 300, as referring to rocket weapons. But careful studies by Gode,[43] Partington,[44] and Winter[45] show that all such speculations are based on either legends or myths rather than on history, or else on incorrect translations of the Sanskrit. Hence, they should be discarded. A detailed discussion on this matter is given by Winter.[46]

In tracing the development of firearms in Indian history, P. K. Gode, the Indian scholar, finds that the use of gunpowder in India cannot have begun earlier than the 15th century. Meanwhile, Winter finds that the earliest Indian war rockets appeared during the time of the reign of the Mogul Emperor Akbar (1556-1605), and that they were employed in such frequency and scale that the Indians virtually held the lead in rocket warfare for several hundred years from that time. A series of datable historical accounts since 1565 are cited by Winter,[47] and they are digested here in the appended chronicle. However, not many accounts have been found on gunpowder development in India prior to that time. Thus, accepting these well founded conclusions by Gode and Winter, it is still mysterious how the rocket age could come to India all of a sudden in the 16th century without considerable accounts of previous gunpowder development, which is so vital to the making of a rocket. A reasonable explanation is that gunpowder/rocket technology was brought to India from outside sometime before the 16th century, and that such transference was either direct from China, or through the Moslem countries.

The indirect transference through Moslem countries such as Arabia and Persia is very plausible, since they were both so close to India. As for direct transference, we note that, although China and India are separated by the high Himalayan mountain range, the two countries nevertheless have a long history of contact, comparable to that between China and the West Region. However, such contacts seem to have been mainly on the religious side. As we know, Indian Buddhism was introduced to China in the first century during the reign of Emperor Ming Ti 明帝 of the later Han dynasty. From then on, numerous Indian monks came to China to preach, and many Chinese monks made pilgrimages to India. This long succession of Buddhist interchange between the two countries went on almost continually for many centuries, but such religious contact was not likely to produce the transference of military technology. Besides, diplimatic relations between the two countries were not close, and military contacts were extremely rare. Throughout the period from the -2nd to the 17th century which is of our concern at present, historical accounts show that there were only three military contacts between China and India. The first occurred in 648, when Wang Hsüan-Ts'e 王玄策, the envoy sent by Emperor T'ai Tsung 太宗 of the T'ang dynasty to the Kingdom of Central India, subdued that country by force, and captured its ruler.[48] The second occurred during the first Mongolian westward invasion, when Genghis Khan's army reached the Indus River in 1223; but his goal was Khwarizm, not India, and he soon turned back on his triumphal return after the former was conquered.[49] The third occured during the expedition of Cheng Ho 鄭和, the eunuch admiral sent by Emperor Ch'eng Tsu 成祖 of the Ming dynasty, when his fleet subdued Ceylong and captured its king in 1408.[50] All three of these were limited wars of small scale and short duration,[51] hence they do not seem to have produced significant technological transference. However, India and China had maintained their commercial contacts almost continually for many centuries. As India was not on the main silk road, her communication with China was mainly along maritime routes. It is said that Indian ships regularly visited Chinese sea ports as early as the 2nd century, and at the end of the 4th century, Chinese ships reached Ceylon, according to Needham.[52] Such sea trade routes were at an ever prosperous stage from the 13th century onward, when China had entered her gunpowder age. During the Yuan dyasty, Indian merchant ships quite frequently visited Chuangchow, one of the greatest Chinese sea ports at that time, and at the same time Chinese merchant ships reached Malarbar on the southwestern coast of India, according to Marco Polo.[53] Such prosperous sea trade routes continued to the Ming dynasty. Thus the direct transference of gunpowder/rocket technology to India from China between the 13th century and the 16th century through such maritime routes was quite possible.

The above brief review shows that, in comparison with the Moslem countries, India was in a less favorable position for such technological transference. This is probably the reason that, while gunpowder/rocket technology was quickly absorbed by the Moslem countries soon after the gunpowder age came to China, India was relatively late in such development. In fact, when India began to show activity in rocket weapons in the 16th century, such technology had already been known to the Europeans for about 300 years. The comparatively late development of gunpowder/rocket technology in India was probably brought about mostly by indirect transference through her Moslem neighbors, and to a lesser degree by direct transference from China. The interchange of such technology between India and Europe in the medieval ages is possible, but evidence has yet to be found. In any case, the contribution of India to the transference of gunpowder/rocket technology from China to Europe, if any, seems much less than that of the Moslems.

D. Conclusions

From the foregoing considerations it seems that it was mainly through the Moslem countries that Europe obtained gunpowder/rocket technology from China. In fact, it was believed that Roger Bacon and Albertus Magnus, the two European pioneers in such technology, utilized Arabic works.[54] It is interesting to note here the fact that Bacon presented his gunpowder formula in the mysterious form of an anagram-like cipher, which became understandable only after proper deciphering; and such was the usual practice of Chinese Taoists in presenting their alchemical findings, though in a slightly different manner. For example, in Wei Po-Yang's Ts'an T'ung Chi, there is a poetic passage[55]:

> A young lady on the river is mysterious and divine;
> When fire comes, she flies away without a trace of dust.
> Vanishing like a spirit, and secreting like a dragon,
> Nobody knows to where she retreats.
> How to put her under control?
> The use of the yellow-bud is suggested.

Here the young lady means mercury, the yellow-bud, sulphur, and the whole passage states the fact that the heating of mercury with sulphur will make the former subdued. Thus, it actually implies the chemical formula $Hg + S = HgS$, according to Li.[56] Such examples are numerous in the Tao Tsang. Whether Bacon's way of presenting his gunpowder formula was inspired by the practice of the Chinese Taoists we do not know, but such similarity is something worth our attention.

Historical accounts[57] show that in the two earliest campaigns in which European rockets were in action, namely the battle of the Iberian Peninsula in 1249, and the battle of Valencia in 1249, those who employed the rockets were both Moslems. This is further evidence showing the Moslem connection of the early European rockets.

Summarizing the materials presented above, we may draw our conclusions as follows:

1. The westward transference of gunpowder/rocket technology from China to Europe was to a large extent through the Moslem countries in the 13th century and the three massive westward invasions by the Mongols during that period. The establishment of the trans-Asia land road under the great Mongolian Empire facilitated such transference.
2. Such technology was transferred to India from China at a somewhat later time, possibly through direct contacts, but more likely through the Moslem countries. The eventual transference to Europe with India as the intermediary, if any, though possible, was not significant in comparison with the Moslems.

IV. The Eastward Transference

All countries in Eastern Asia have been within the sphere of influence of China. Those closest to China are Korea and Japan in the Northeast, and the countries of Indochina in the Southeast. In the following the transference gunpowder/rocket technology from China to these countries will be discussed.

A. Korea

Historically, Korea, with her proximity to China's northeastern border has been closely associated with China culturally and politically since many centuries B.C. During such association, Korea was sometimes made a part of China as a province, sometimes independent, but mostly a tributary state, all the way to the 19th century.

The history of Korea shows that the first government office in charge of firearms was established in 1377 on the recommendation of Choi Mu-Son 崔茂宣, a Korean specialist.[58] This was about 300 years after such an office was established in China during the Northern Sung dynasty[59] and nine years after Chu Yüan-Chang 朱元璋 overthrew the Mongol government of the Yuan dynasty and established the Ming dynasty. Choi was known to have learned gunpowder technology from a man named Li Yuan 李元, who used to supply saltpeter for the Yuan army when Korea was still under Mongolian occupation before the downfall of the Yuan dynasty. Since 1377, Choi had made various kinds of gunpowder firearms, including fire arrows which are believed to be of the gunpowder-charged incendiary type like those used by the Chinese in the earlier years of the Sung dynasty. The truly reaction propelled Korean rocket was also made by Choi sometime between 1377 and 1389, and it was called 'ju-hua' (literally, 'running fire') according to Chae.[60] Such running-fire rockets were once the favorites of the Korean army, but after 1448 they were replaced by those of a new generation, called 'magic arrows,' which were of four different kinds; small, medium, large, and multiple. Descriptions of these magic arrows, including drawings with dimensions are found in Chae.[61] What is worth our particular attention here is that 'magic arrow' was the name of a popular kind of Chinese rocket, which had been listed as one of the standard armaments for the government forces in the earlier years of the Ming dynasty, about

65 years before its emergence in Korea.[62] They were written with the exact same Chinese characters 神機箭 and pronounced alike ('shen chi chien' in Chinese; and 'sin-gi-chian' in Korean). Furthermore, by the year 1448, multiple rockets had also been well developed and employed in military actions in China for nearly a half-century.[63] Thus it is obvious that these magic arrows made by Choi must have been a Korean version of the Chinese model. In fact, many of the kinds of firearms Choi made had names either identical or similar to those found in China, such as 'huo t'ung' 火熥 ('fire barrel'), 'huo p'ao' 火砲 ('fire catapult'),[64] etc. Later during the two invasions by the Japanese warlord, Tyotomi Hideyoshi 豐臣秀吉 in the years 1592 to 1598, the use of many kinds of gunpowder firearms (including magic arrows) by the allied fleets of Korea and China was reported.[65]

The preceding account shows that Chinese gunpowder/rocket technology was first transferred to Korea in the early years of the 14th century through the Mongolian occupation army, and that such transference continued from then on through the close relationship between Korea and China in the Ming dynasty, and joint military operations such as those mentioned above must have promoted such transference.

B. Japan

Japan also has a long history of contact with China, traceable to at least several centures B.C. But, unlike Korea, she is geographically separated from the Asian mainland by the sea, and hence never became a subordinate of China, except nominally for some short periods, even when China was under the aggressive Mongol government in the Yuan dynasty. With Korea interposed between the Japanese Islands and the Chinese northeastern coast, she acquired Chinese culture essentially through Korea in the early years of the

Earlier Han dynasty. But later, direct contacts were gradually established, diplomatic envoys exchanged, sea trade routes opened, and numerous Japanes scholars and students were sent to China for study, and occasionally there were wars between the two countries.

After China entered her gunpowder age, the first gunpowder firearm to fall on Japanese soil was brought by the Mongol invasions during the battles on the Kyushu coast in the years 1274 to 1281. A detailed description of such firearms is found in the Japanese account, Hachiman Gudoki,[66] from which the following is a passage rendered in English by Arakawa[67]:

> When they (the Mongols) withdrew, they set off so-called Teppo (literally, gun) which lets off violently projectiles made of iron by virtue of powderlike mixtures. When these projectiles beat against the land and break off, fire runs all around, smoke makes darkness, and sound is very noisy. Their noise and bright glare have thrown us into confusion, startling, dazzling and bewilderment. Many soldiers were killed on the field in consequence of these new weapons.

Pictorial illustrations are shown in a later work, the Moko-Shurai-Ekotoba (literally, The Mongolian-Invasion Pictorial).[68] Just what was this Teppo? Arakawa regards it as a naive rocket. But it was more likely a sort of grenade such as the 'heaven-shaking thunder' employed by the Chin Tartars against the Mongols in the battle of K'aifeng-fu some forty years earlier, according to Wang.[69] Whatever its constructional details, it was no doubt a gunpowder firearm. By the great shock produced on the Japanese people as described in the Japanese account, it is evident that they had no knowledge of gunpowder or gunpowder firearms at that time. However, a hundred years after the Mongol invasion, there is a record in Chinese history, which says:

> A monk, named Ju-Yao 如瑤, was sent to China by the Japanese ruler in 1382 during the reign of the first emperor T'ai Tsu (Chu Yüan-Chang) of the Ming dynasty. Inside the huge candles which were to be presented by Ju-Yao to the Emperor, there was found concealed gunpowder, together with some swords and knives. It was believed to be a plot aiming at the assassination of the Emperor. This plot was not actually carried out, however.[70]

This story shows that there was gunpowder in Japanese hands at that time. Whether it was made in Japan or obtained from China we don't know.

It is strange that not much has been found about the development of gunpowder/rocket technology in Japan afterwards that parallels the development in Korea, despite the many occasions in which the Japanese encountered various gunpowder firearms in their fights with Korean and Chinese from the 14th to the 16th centuries. It seems that instead of Chinese firearms the main Japanese interest was in the bird-beak guns first imported from Portugal in 1543. Historical accounts show that the Japanese engaged in its imitation by their own technicians, and actually put such guns to great use during their invasions of Korea under Hideyoshi in 1592-98. This was obviously due to the greater destructive power of such Western weapons in comparison with the Chinese gunpowder firearms. To show how earnest the Japanese were in learning the making of such Portuguese weapons, a story says that the man who was in charge of imitating the bird-beak gun sent a young maiden 17 years of age to the Portuguese smith to please him in order to get all the secrets of such technology.[71] This happened around the middle of the 16th century, which was the time when Western technology began its penetration into the East, hence the Japanese no longer imitated the Chinese model as they had traditionally done, but turned to the West instead. In fact, during the same period, China was also considerably influenced by Western technology, and the bird-beak guns and barrel cannons from Europe were also her favorites. However, in view of the close Japanese contacts with the Chinese and Koreans, if they had some knowledge of gunpowder/rocket technology prior to the middle of the 16th century, it is likely that they learned it either directly from China, or indirectly through Korea.

C. Countries in Indochina

Like Korea and Japan, the countries in Southeast Asia were also under the great influence of China for over a millennium. In particular, those in Indochina, such as Burma, Thailand, Laos, Cambodia and Vietnam, which are so close to the southwestern border of China, were made either Chinese provinces, or tributary states in their long association with China. However, the gunpowder/rocket technology which spread to those countries seems to have been mainly pyrotechnics for festival purposes. A lively description of the firework activities in Chen-La 眞臘 (modern Cambodia) in the 13th century is found in Chou Ta-Kuan's 周達觀 book, Chen-La Feng-T'u Chi《眞臘風土記》(Life and Customs in Chen-La), a written account of what he saw and heard during his one-year stay in that country when he was an officer under the special envoy sent by the Chinese government to Chen-La during 1296-97. A passage of this book reads[72]:

> In the first month of the year, corresponding to the 10th month in the Chinese calendar, a hugh booth is set up in the front of the palace, capable of accommodating a thousand people, and decorated with lanterns and flowers. Across the river, tall booths made by fastening together wooden bars like those temporary structures used in building pagodas are erected on the bank. Each has a height of about 20 'chang' 丈 (about 61 meters). Every night, two to three, or five to six sets of fireworks and firecrackers are set on these booths, and ignited in the presence of the king. The fireworks can be seen from a hundred 'li' 里 (about 55 km) away, and the firecrackers, each the size of a gun, produce a noise which shakes the whole city.

The size and performance as described above might be inaccurate and a little exaggerated, but the display was no doubt on a grand scale. This accout shows that the pyrotechnics in Chen-La had really reached an advanced stage at that time. Also found in Chou's book, the following information is noteworthy: 1) sulphur and saltpeter (the smoke-producing kind) were among the Chinese merchandise that the Chen-La people like best; and 2) the soldiers of the Chen-La army had only javelins and shields as their weapons, but no bows and arrows, not catapults. These two facts together indicate clearly that the sulphur and saltpeter imported from China were used to manufacture the gunpowder for their fireworks and firecrackers, not for war rockets.

Such firework activities were also found in Burma, Thailand, Laos, and Vietnam, according to Winter.[73] He himself even witnessed such a splendid display during the so-called annual Boun Bang

Fei Rocket Festival in modern Thailand in 1977. The use of signal rockets by the Siamese in their fight against the Cambodians in 1593, and that of war rockets by the Vietnamese in their fight against the French in 1883 are also reported in Winter's article,[74] but no details are given.

As to the source of such technology, pyrotechnics or rocketry, China was of course a principal one, supplying the technology as well as some raw materials mainly through diplomatic and commercial channels from the 13th century on. However, the Indian influence cannot be overlooked, since India is also geographically proximal to the countries in Southeast Asia and historically has had close contacts with them traceable at least to the early centuries A.D. In fact, the cultural influence of India can be felt in most parts of Southeast Asia, not only in Indochina. In particular, the Indians are fond of festival rockets, and war rockets had been well developed and used in India since the 16th century. By focusing attention on the size of the festival rocket, Winter finds that the huge designs of those found in the countries in Indochina are more like those found in India, which were usually large, but unlike those found in ancient China which were usually small. Such a comparison makes Winter think that it was India which supplied such technology to these countries, rather than China.[75] With due consideration to Winter's speculation, it seems reasonable to say that gunpowder/rocket technology, mainly the pyrotechnics, was transferred from China to Indochina from the 13th century on, and later it was further developed in that region under the great Indian influence.

The transference of gunpowder/rocket technology from China to other countries in Southeast Asia beyond Indochina is left for future study.

V. Final Conclusions: A Summary

Based on the historical review presented above, the questions posed in the Introduction may now be answered as follows:

1. The development of gunpowder/rocket technology in China, which evolved from the art of alchemy, was independent from its early beginning in the -2nd century to the middle of the 16th century.
2. China entered her gunpowder age in the early 11th century. Festival rockets had become popular by the 12th century, but war rockets did not appear until the 13th century during the Battle of K'aifeng-fu in 1232.
3. The 13th century was an epochal period for gunpowder/rocket development: before it no datable events on such technology are found in the world outside China, but from this century onward such events turned out in succession in the other parts of Asia, and in Europe.
4. The possible routes for the westward transference of such technology from China to Europe were mainly through the Moslem countries from the 11th to the 13th centuries; and the Mongols, during their establishment of the great Mongol empire comprising the entirety of the Asian mainland and a part of Europe, played an important role in such transference, especially of gunpowder firearms. The role played by India in such transference was minor.
5. In the northeast direction such technology was transferred mainly to Korea during the 14th to the 16th centuries through her direct contacts with China, and what was transferred most was firearms and war rockets. Some spreading to Japan in the same period directly from China or indirectly through Korea is likely.

6. In the southeast direction such technology spread to the countries of Indochina, such as Burma, Thailand, Laos, Cambodia, and Vietnam through their direct contacts with China from the 14th century onward, and such transference was mainly in pyrotechnics, but the later development in this region was significantly influenced by India.
7. After the middle of the 16th century, Chinese technology was greatly influenced by the West, the traditional Chinese gunpowder firearms and war rockets were gradually replaced by Western guns and cannons, and China no longer provided the source of such technology, leaving only pyrotechnics still flowering.

A tentative flow chart depicting the routes of transfer prior to the 16th century is shown in the accompanying pages.

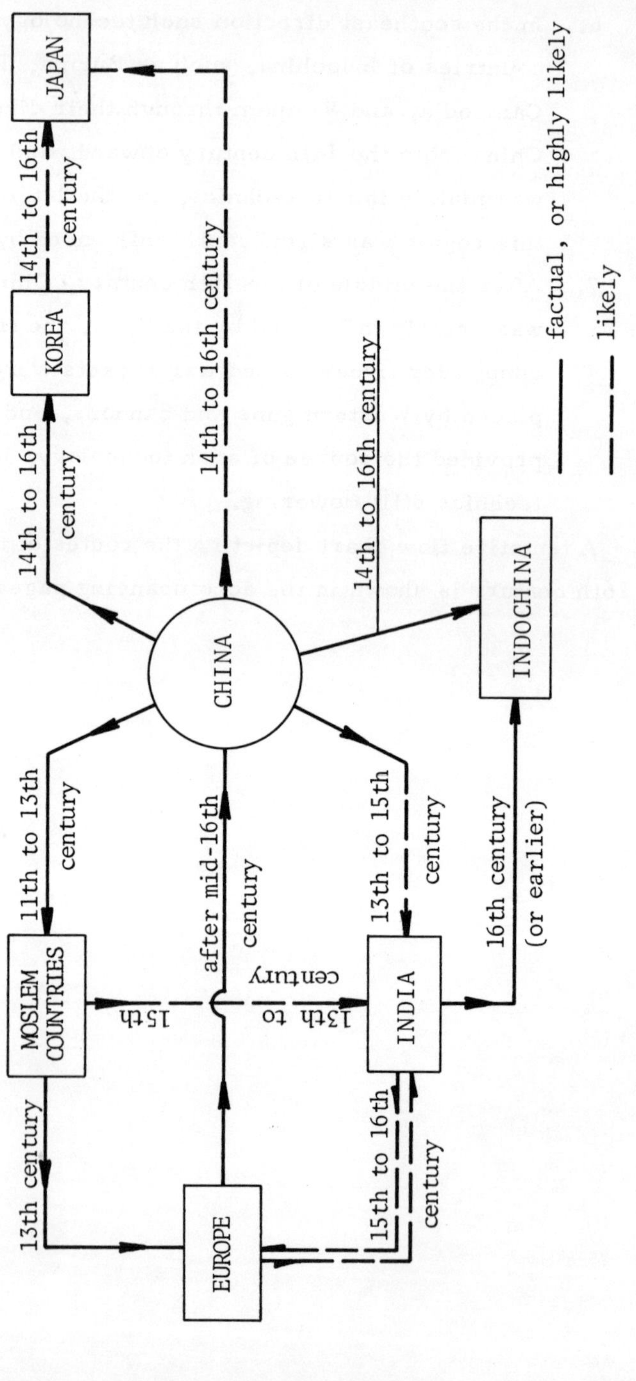

A Tentative Flow Chart for the Transference of Gunpowder/Rocket Technology (11th to 16th century)

References

1. Wu Pu 吳普, Shen Nung Pen Ts'ao Ching《神農本草經》(Classical Pharmacapoeia of the Heavenly Husbandman) 2nd century; volume 1, pp. 5-6; volume 2, pp. 2-3.
2. Liu An 劉安, Huai Nan Tzu 《淮南子》(The Book of the Prince of Huai Nan), c. -120, volume 3, Astronomy.
3. Wei Po-Yang 魏伯陽, Chou-I Ts'an T'ung Ch'i《周易叅同契》(The Kinship of the Three and the Book of Change) of +142; section 40.
4. Chen Kuo-Fu 陳國符, Tao-Ts'ang Yüan-Liu Hsü-K'ao《道藏源流續考》(Continued Research on the History of the Taoist Patrologies)(Ming-Wen Book Co., Taipei, 1983), chapter 3, p. 396.
5. Wu Pu, op. cit. (Ref. 1), volumes 1 and 2.
6. Chang Chung-Ching 張仲景, Chin-Kuei Yao-Lüeh 《金匱要畧》(The Essentials in the Golden Chest) + 180; volume 2.
7. Nathan Sivin, Chinese Alchemy: Preliminary Studies (Harvard University Press, Cambridge, Massachusetts, 1968), Chapter IV.
8. Fan Yeh 范曄, Hou Han Shu《後漢書》(History of the Later Han Dynasty) 450; volume 15.
9. For the pros, see for example, Feng Chia-Sheng 馮家昇 Huo-Yao Chi Ch'i Hsi-Ch'uan《火藥及其西傳》(Origin of Gunpowder and Its Western Spread) (Shanghai, 1954), p. 41; Joseph Needham, 'The Epic of Gunpowder and Firearms, Developing from Alchemy,' Science in Traditional China: A Comparative Perspective (Linking Publ. Co., Taipei, 1982), p. 29; Lung Ts'un-Ni 龍村倪, 'Fu-Huo-Liu-Huang-Fa Ying Kui Sun Chen Jen, Yi Chang Yi Chih Tsui Ku Huo-Yao Hsiang Fang Pien' 〈伏火硫黃法應歸孫眞人，一張已知最古火

藥詳方辨〉 ('On the Earliest Known Gunpowder Recipe, A Method of Subduing Sulphur by Fire, by Sun Chen-Jen'), Science Monthly (Taipei) 13 (No. 11), 64-70 (1982). For the cons, see Nathan Sivin, op. cit. (Ref. 7), pp. 76-77; Liu Kwang-Ting 劉廣定, 'Tsai-T'an Fu-Hui-Liu-Huang-Fa ti Tso-Che Wen-T'i' 〈再談伏火硫黃法的作者問題〉 ('Rediscussion on the Author of the Method of Subduing Sulphur by Fire'), Science Monthly (Taipei), 13 (No. 11), pp. 70-72 (1982).

10. Chen Kuo-Fu, op. cit. (Ref. 4), chapter 2, p. 337.

11. Weng Tung-Wen 翁同文, 'Chen Yuan Miao Tao Yao-Lüeh ti Ch'eng-Shu Shih-Tai Chi Hsiang Kuan ti Huo-Yao-Shih Wen-T'i' 〈眞元妙道要畧的成書時代及相關的火藥史問題〉 ('The Dating of Chen Yüan Miao-Tao Yao Lüeh and Gunpowder') Nanyang University Journal, No. 5, pp. 33-80 (1972).

12. Sun Fang Toh 孫方鐸, 'Rockets and Rocket Propulsion Devices in Ancient China' (14th IAA International Symposium on the History of Astronautics, 31st IAF Congress, Tokyo, 1980) Journal of Astronautical Sciences, Volume XXIX, Number 3, pp. 289-305 (1981).

13. Wang Ling 王鈴, 'On the Invention and Use of Gunpowder and Firearms in China,' ISIS, 37, 160-178 (1947).

14. Sun Fang Toh, op. cit. (Ref. 12); Sun Fang-Toh, 'On the Early Rocket Weapons in China' (18th IAA International Symposium on the History of Astronautics, 35th IAF Congress, Laussanne, Switzerland, 1984), Paper No. IAA-84-258.

15. For details, see T'o-T'o 脫脫 and Ouyang Hsüan 歐陽玄, Chin Shih 《金史》(History of the Chin Dynasty), 1345, volume 113; Tu Chi 屠寄 Meng-Wu-Erh Shih-Chi 《蒙兀兒史記》(History of the Mongols), 1934, volume 29.

16. The other one being the so-called 'cheng tien rai' 震天雷 ('heaven-shaking thunder') which was a form of grenade.

17. 1 pace $\approx$ 5 feet.

18. Sun Fang-Toh, 'On the Early Rocket Weapons in China,' op. cit. (Ref. 14).

19. Ming Hui Tien 《明會典》(Collection of the Archives of the Ming Dynasty), compiled by Imperial Order, volume 156, 1587.

20. Multi-rockets, 'yi wo feng' 一窩蜂 ('beehives') were used, see Chang Hsüan 張萱, Hsi-Yüan Wen-Chien-Lu 《西園聞見錄》(Seen and Heard in the West Garden), volume 73, 1582.

21. Multi-rockets, 'chiu lung chien' 九龍箭 ('nine dragon arrows') were used, see Chang T'ing-Yü 張廷玉, Ming Shih 《明史》(History of the Ming Dynasty), 1739, edited by Yang Chia-Lu 楊家駱 (Ting-Wen Book Company, Taipei, 1975), volume 92.

22. The particular kinds of rockets used is unknown, see Tien Ju-Ch'eng 田汝成, Yen-Chao Chi Wen 《炎徼記聞》(Memories on the Southern Border), 1558.

23. Chao Tieh-Han 趙鐵寒, The Invention of Gunpowder (National Institute for Compilation and Translation, Taipei, 1978).

24. Chou Mi 周密, Wu-Ling Chiu-Shih 《武林舊事》(Old Anecdotes of Wu-Ling), 1269.

25. Joseph Needham, 'The Epic of Gunpowder and Firearms: Developing from Alchemy,' Science in Traditional China: A Comparative Perspective (Linking Publ. Co., Taipei, 1982), p. 47.

26. See, for example, Willy Ley, Rockets, Missiles, and Men in Space (Viking Press, New York, 1968), pp. 53-54.

27. See, for example, E. M. Emme, A History of Space Flight (Holt, Rinehart and Winston, New York, 1965), p. 64.

28. Berthold Laufer, Sino-Iranica: The Chinese Contribution to the History of Civilization in Ancient Iran (Field Museum of Natural History Publications, 1914), Anthropological Series, volume 15, number 31. Chinese translation by Tu Cheng-Sheng 杜正勝 (Chung Hua Book Co., Taipei, 1975), p. 443 (Chinese translation).

29. Ley, op. cit. (Ref. 26), p. 53.

30. Feng Chia-Sheng, Huo-Yao Chi Ch'i Hsi-Ch'uan, op. cit. (Ref. 9), p. 46; Ley, op. cit. (Ref. 26), p. 53.

31. Feng, op. cit. (Ref. 9), p. 53; Needham, 'The Epic of Gunpowder and Firearms,' op. cit. (Ref. 25), p. 40.

32. Feng, loc. cit. (Ref. 9).

33. Pan Ku 班固, Ch'ien Han Shu 《前漢書》 (History of the Earlier Han Dynasty), c. 100; volume 61.

34. Fan Yeh 范曄, Hou Han Shu op. cit. (Ref. 8), volume 77.

35. Joseph Needham, Science and Civlisation in China, volume I, chapter 7, Cambridge University Press, 1954.

36. Feng Chia-Sheng, op. cit. (Ref. 9), p. 64; Needham, 'The Epic of Gunpowder and Firearms,' op. cit. (Ref. 25), p. 49.

37. Tu Chi, op. cit. (Ref. 15), volume 59.

38. Tu Chi, op. cit. (Ref. 15), volume 144, where the footnote says, " 'huo kuan' was interpreted by Tu Shao Shih 屠孝寔 (the son of the author, Tu Chi 屠寄) as a jet engine with burning oil." This interpretation is in agreement with that given in the Wu-Ching Tsung Yao 《武經總要》 (Collection of the Most Important Military Techniques) by Tseng Kung-Liang 曾公亮 and Ting Tu 丁度, of 1044; volume 12, pp. 66-69.

39. D'Hosson, Histoire des Mongols (Amsterdam, 1852), 4 volumes; Chinese translation by Feng Chen-Chün 馮承鈞 (Commercial Press, Taipei, 1934); volume 2, chapter 3, p. 228 (Chinese translation).

40. Feng Chia-Sheng, op. cit. (Ref. 9), p. 63.

41. Ibid., p. 51.

42. Joseph Needham, Science and Civilisation in China (Cambridge University Press, Cambridge, 1954), volume I, p. 188.

43. P. K. Gode, Use of Guns and Gunpowder in India from A.D. 1400 Onwards (Karnatak Publ. House, 1939), A Volume of Indian and Iranian Studies, p. 20.

44. J. R. Partington, A History of Greek Fire and Gunpowder (W. Heffer and Sons, Cambridge, 1960), p. 211.

45. F. H. Winter, 'Rocketry in India from Earliest Times to the 19th Century,' Journal of British Interplanetary Science, 32, 467-471 (1979).

46. Ibid.

47. Ibid.

48. Liu Hsu 劉昫, Chiu Tang Shu《舊唐書》(Old History of the T'ang Dynasty), 945, volume 198.

49. Tu Chi, op. cit. (Ref. 15), volume 144, pp. 12-14; also Chang Hsing-Lang 張星烺, Chung Hsi Chao-Tung Shih-Liao Hui-Pien《中西交通史料彙編》(The Materials for a History of Sino-Foreign Relations) (World Book Co., Taipei, 1962), volume 6, pp. 465-467.

50. Chang T'ing-Yu, Ming Shih, op. cit. (Ref. 21), volume 304, pp. 7766-67.

51. In addition to these three, an expedition sent to India by Mangu Khan in 1253 is recorded in history (see Chang Hsing-Lang, op. cit. (Ref. 49), volume 6, p. 6), but nothing is recorded about what happened afterwards.

52. See Needham, Science and Civilisation in China, op. cit. (Ref. 42), volume I, p. 179.

53. Henry Yule, The Book of Ser Marco Polo, the Venetian (Murray, London, 1903). Chinese translation by Chang Hsing-Lang 張星烺 (Commercial Press, Taipei, 1965); pp. 336-337, 424-426 (Chinese translation).

54. Feng Chia-Sheng, op. cit. (Ref. 9), pp. 70-73; F. H. Winter, 'A New Look at Early Chinese Rocketry, 1200-1900,' Journal of the British Interplanetary Society, 35, 522-529 (1984).

55. Wei Po-Yang, op. cit. (Ref. 3), Chapter 72.

56. Li Ch'iao-P'ing 李喬萍, Chung-Kuo Huo-Hsüeh Shih《中國化學史》(History of Chemistry in China) (Commercial Press, Taipei, 1975), p. 32.

57. W. R. Maxwell, 'The Early History of Rockets,' Journal of the British Interplanetary Society, 35, 176-184 (1982).

58. Y. S. Chae, 'A Study of Early Korean Rockets (1377-1600),' Journal of the British Interplanetary Society, 38, 169-178 (1985), or Wang Ling, op. cit. (Ref. 13), p. 169.

59. Fung Chia-Sheng, op. cit. (Ref. 9), p. 20.

60. Y. S. Chae, op. cit. (Ref. 58).

61. Ibid.

62. Ming Hui Tien, op. cit., volume 156.

63. Sun Fang-Toh, 'On the Early Rocket Weapons in China,' op. cit. (Ref. 14).

64. Yi Pyong-Do 李丙燾,《韓國史大觀》(History of Korea), Chinese Translation from Korean by Hsü Yü-Ch'eng 許宇成 (Cheng-Chung Book Co., Taipei, 1961), p. 260 (Chinese translation).

65. Ibid., pp. 341-346 (Chinese translation).

66. Written in the late 13th century, author unknown.

67. H. Arakawa, 'Rockets Fired by the Mongol Forces in 1274 A.D., Proceedings of the 2nd International Symposium on Rockets and

Astronautics (Tokyo, 1960), pp. 330-331.

68. A Japanese imperial property, painted in 1293.

69. Wang Kung-Chi 汪公紀, Jih-Pen Shih-Hua, Chung-Ku Pien 《日本史話，中古篇》 (Epic of Japanese History, Medieval Ages), (Linking Publ. Co., Taipei, 1982), p. 111, which is based on A History of Japan, by Kunoda Toshio, 1965, pp. 90-92.

70. Chang T'ing-Yu, op. cit. (Ref. 21), volume 210: Japan.

71. Cheng Hsüeh-Chia 鄭學稼, Jih-Pen Shih 《日本史》 (History of Japan) (Li Ming Book Co., Taipei, 1977), volume 3, pp. 4-5.

72. Chou Ta-Kuan 周達觀, Chen-La Fen-T'u Chi 《眞臘風土記》 (Life and Customs in Chen-La [Cambodia]), 1297, p. 11.

73. F. H. Winter, 'A New Look at Early Chinese Rocketry, 1200-1900,' op. cit. (Ref. 45).

74. Ibid.

75. Ibid., pp. 525-526.

76. Meng Yao-Fu 孟要甫, Chu-Chia Shen P'in Tan Fa 《諸家神品丹法》 (Methods of Various Schools for Magical Elixir Preparations), 10th century.

77. Chao Nai-An 趙耐安, Ch'ien Hung Chia Kung Chih Pao Chi Ch'eng 《鉛汞甲庚至寶集成》 (Complete Compendium on the Perfected Treasure of Lead, Mercury, Wood, and Metal), 808.

78. Chen-Yüan Miao-Tao Yao-Lüeh 《眞元妙道要略》 (Classified Essentials of the Mysterious Tao of the True Origin), attributed to Cheng Shih-Yüan 鄭思遠, 3rd century.

79. Sung Ming-Chiu 宋敏求, Tung Ching Chih 《東京志》 (About the Eastern Capital), early 11th century.

80. Shih Liang-Mou 石良茂, Pei Jung Yeh-Hua 《避戎夜話》 (Some Night Talks During My Hiding From War), 1130; Hsu

Meng-Hsin 徐夢莘, San-Tsao Pei Ming Huai Pien《三朝北盟會編》(Collection of Documents on the Treaties with the North in Three Reigns), 1194.

81. T'o-T'o 脫脫, Chin Shih 《金史》 (History of the Chin Dynasty), 1345, volume 77.

82. Yang Wan-Li 楊萬里, Hai-Chiu Fu Hou-Hsu, Ch'eng-Chai Chi《海鰌賦後序，誠齋集》(Epilogue to the Ode of the Sea Eel, The Collected Work of Yang Ch'eng-Chai), volume 44, 1170; J. Pan, 'On the Origin of Rockets,' Nature 3, 173-184 (1984).

83. Chou Mi 周密, Chi-Tung Yeh-Yü 《齊東野語》(Country Talks of Eastern Chi), volume 11, 1269.

84. Liu Ming-Chung 劉敏中, Ping Sung Lu 《平宋錄》 (The Conqueror of Sung), 1304.

85. Ch'i Chi-Kuang 戚繼光, Chi-Hsiao Hsin Shu《紀效新書》 (A New Treatise on Military and Naval Efficiency), volumes 1, 6, and 15, 1575.

86. Mao Yüan-I 茅元儀, Wu Pei Chih 《武備志》(Treatise on Armament Technology), volumes 126-129, 1628.

87. Sung Yin-Hsing 宋應星, T'ien-Kung K'ai-Wu《天工開物》(The Exploitation of the Work of Nature), 1637, volume 15.

88. Chiao Hsü 焦勗, Tse K'o Lu 《則克錄》(Elements of Fire Warfare), 1643.

Table II. A Preliminary Chronicle of Gunpowder/Rocket Development: 8th - 17th Century

A. Development in Ancient China: 8 - 12th Century

Century	Year	Events
8-9th	760-808	Appearance of proto-gunpowder recipes. 1. A formula for subduing sulphur by fire is possibly contributed by Sun Ssu-Mo.[a] 2. A formula for subduing alum by fire is contributed by Chin-Hua Tze (808).[b]
10th	950	Accidental deflagration of gunpowder-like mixture recorded.[c]
	970	Gunpowder fire arrows presented to Emperor Sung Tai-Tsu by Officer Feng Chih-Sen, tested, and awarded.[d]
	975	Governor of Wu-Yüeh sends to Emperor Sung Tai-Tsu soldiers trained for gunpowder fire arrows.[e]
	998	Captain of the imperial navy presents to Emperor Sung Tseng-Chung several kinds of firearms - including the fire arrow, the fire ball, and the barbed fire ball - and is awarded.[f]
11th	1044	Three gunpowder formulas for military applications appear in the imperial publication on armaments, including many forms of gunpowder-powered firearms.[g]
	1070	Gunpowder manufacturing factory in the capital, Pienking, is described in Sun Ming-Chiu's book, <u>Tung-Ching Chih</u>.[h]

Table II (continued)

Century	Year	Events
12th	1126	Battle of K'aifeng-Fu (Pienking), between the Sung and the Chin Tartars: both sides use firearms powered by the gunpowder-like mixture, such as the fire arrow, the fire trebuchet, etc.[i]
	1130	Naval engagement in Hwang-tien Dong on the Yang-tze River; the Chin Tartars attack the Sung fleet under the command of General Han Shih-Chung with the gunpowder fire arrows.[j]
	1161	Battle of Tsai-Shih Chi on the Yangtze River: Sung's navy under the command of Yu Yun-Wen defeats the fleet of King Liang of the Chin Tartars with Pi-li Pao, a firearm, possibly a gunpowder-powered rocket-like weapon.[k]
	1165	Fireworks prevail in the region of the Southern Sung: magnificent performances as court entertainment are recorded.[l]

*Remarks on the Chronicle Table:

1. Only the datable events are listed; and the table is by no means exhaustive.
2. Of the various gunpowder firearms, attention is focused on the rocket, with only a few of the other kinds listed as examples. Owing to the great variety of firearms which appeared after the beginning of the gunpowder age, listing all of them would require a separate table, which is beyond the scope of the present paper.

[a] See Ref. 76.
[b] See Ref. 77.
[c] See Ref. 78.
[d]
[d] T'o-T'o (see Ref. 15), volume 197.
[e] Ibid., volume 3.
[f] Ibid., volume 197.
[g] Tseng and Ting (see Ref. 38), volumes 11 and 12.
[h] See Ref. 79.
[i] See Ref. 80.
[j] See Ref. 81.
[k] See Ref. 82.
[l] Chou Mi (see Ref. 24).

Table II (continued)

B. Development in China and the Outer World: 13th Century

Year	China	Arabia, Europe	Eastern Asia
1220-1221		Genghis Khan's invasion of Khwarizm. General Kuo Pao-Yü of the Mongolian army attacks the energy with gunpowder fire arrows (1220).[a] Prince Tuliu attacks the enemy with gunpowder firearms in the campaign of Khroasan (1221).[b]	
1227	A minor accident in the royal party: a gunpowder-powered, reaction propelled device, a 'yen-huo' called 'the ground rat' gets out of control, and frightens the Empress-Dowager by running beneath her throne.[c]		
1232-1233	Emergence of the military rocket in China during the war between the Ghin Tartars and the Mongols: Battle of Pienking (K'aifeng-fu): 'fei huo chiang' is one of the two		

Table II (continued)

Year	China	Arabia, Europe	Eastern Asia
	new firearms used by the garrison force of the Chin Tartars against the Mongols, the other being a form of grenade, called 'Chen-tien rai' (1232).[d] Battle of Kwei-teh: the same 'fei huo chiang' is again used by the Chin Tartars to defeat the Mongols (1233).[e]		
1241		Battle of Leignitz: the Mongolian invaders attack the allied army of Poland and Germany with a kind of gunpowder fire-arm, possibly the smoke ball.[f]	
1247		Roger Bacon (English monk) gives a gunpowder formula in his <u>Epistola</u>, written in Latin.[g]	
1249		Battle of the Iberian Peninsula: the Moors use rockets.[h]	

Table II (continued)

Year	China	Arabia, Europe	Eastern Asia
1258		Battle of Bagada: the invading Mongolian army under Hulagu Khan use gunpowder firearms.[i]	
1274-1281			Mongolian invasion of Japan: a rocket-like firearm is used by the Mongols in the campaigns on Kyushu.[j]
1274-1275	Mongolian invasion of the Sung: 'huo p'ao' (the fire trebuchet) is employed by the Mongols under the command of Premier Pei Yen to defeat the Sung land and naval forces at San Yang and Chang-chow.[k]		
1275		Albertus Magnus (German bishop) gives a gunpowder recipe and directions for making rockets, written in Latin.[l]	

Table II (continued)

Year	China	Arabia, Europe	Eastern Asia
1280		Hassan er-Rammah (Arabian) gave recipes for making gunpowder and directions for making rockets in his book, Of Fighting on Horseback and with War Engines.[m]	
1288		Battle of Valencia: rockets are used by the Arabs.[n]	
1296			Hugh fireworks in Cambodia are reported.[o]

[a]Tu Chi (see Ref. 15), volume 59.
[b]Ibid., volume 144.
[c]See Ref. 83.
[d]T'o-T'o, Chin Shih (see Ref. 81), volume 113.
[e]Ibid., volume 116.
[f]D'Hosson (see Ref. 39), volume 2, Chapter 3, p. 228 (Chinese translation).
[g]Ley (see Ref. 26), pp. 53-54.
[h]Emme (see Reg. 27), p. 64.
[i]Tu Chi (see Ref. 15), volume 3; Feng (see Ref. 9), p. 51.
[j]Arakawa (see Ref. 67); Wang (see Ref. 69).
[k]See Ref. 84.
[l]Maxwell (see Ref. 57).
[m]Ley (see Ref. 26), p. 53.
[n]Emme (see Ref. 27), p. 67.
[o]Chou (see Ref. 72), p. 12.

Table II (continued)

C. Development in China and the Outer World: 14th Century

Year	China	Europe	Eastern Asia
1326-1345		Gunpowder firearms appear in England and France.[a]	
			Korea: incendiary fire arrows are made by Choi Mu-Son.[b] General Bureau of Gunpowder Firearms is established by the Korean government.[c]
1379		Venetian-Genovese War: rockets are used in the Battle of Chiozza (Italy).[d]	
1383			Korean Navy defeats Japanese invaders by burning their ships with fire barrels and fire trebuchets.[e]
1393	Incendiary fire arrows and 'shen-chi chien' (magic arrows) are listed among standard armaments by the government under		

Table II (continued)

Year	China	Europe	Eastern Asia
	Emperor Ming Tai Tsu, the founder of the Ming dynasty.[f]		
1400	In the campaign of Jingnan, government forces use the multi-rockets 'yi wo feng' ('beehives') against Prince Yen's rebelling army on the Baigou River.[g]		

[a] Feng (see Ref. 9), pp. 74-75.
[b] Chae (see Ref. 58).
[c] Ibid.
[d] Emme (see Ref. 27), p. 66.
[e] Yi (see Ref. 64), p. 250.
[f] Ming-Hui Tien (see Ref. 19), volume 156.
[g] Chang Hsuan (see Ref. 20).

D. Development in China and the Outer World: 15th Century

1405		Konrad Kyeser von Eichstadt (German military engineer) describes several types of rockets in his book, Balli-fortis.[a]	
1420		Joanes de Fontaua (Italian military engineer) sketches various designs for military	

Table II (continued)

Year	China	Europe	Eastern Asia
		rockets, including a rocket torpedo.[b]	
1451			Korean king, Mu Jong designs the fire cart, a multi-rocket launcher.[c]
1464	In the campaign of Luchuan, government forces use the 'chiu-lung chien' ('nine dragon arrow') to pacify the rebelling Miaos, a tribe in Southwestern China.[d]		
1465	In the campaign of Ta-Teng Hsia (Gorge of Great Rattan) government forces use rockets to pacify the rebelling Yaos, a tribe in the mountainous region of Kuanghsi.[e]		
1488			Koreans use a new type of rocket, the 'shen-chi chien,' to displace the old type, the 'running fire.'[f]

[a] Ley (see Ref. 26), pp. 55-56.
[b] Ibid.
[c] Chae (see Ref. 58)
[d] Chang T'ing-Yu (see Ref. 21).
[e] Tien (see Ref. 22).
[f] Chae, op. cit. (Ref. 58).

Table II (continued)

E. Development in China and the Outer World: 16th Century

Year	China	India, Europe	Eastern Asia
1557		Leonhart Fronsperger (German) describes the 'roget,' a firework for amusement, in his book on artillery and fire-arms.[a]	
1565		Battle of Talikota (India): rockets are fired by Vijaynagar army.[b]	
1572		England: festive fireworks display, consisting of flights of rockets, is given in Temple Fields by the Earl of Warwick, Master General of Ordnance, on the occasion of Queen Elizabeth I's visit.[c] Akbar expedition to Gujerat: the enemy's war elephants, frightened by the great noise of the rockets, fly into a panic, causing a general riot.[d]	

Table II (continued)

Year	China	India, Europe	Eastern Asia
	General Ch'i Chi-Kuang trains soldiers for rocket testing and firing.[e]		
1589		Battle of Penconda: Hindu Raja of Aununtgeer's force surrounds the Muslims with a force containing ten thousand rocketeers.[f]	
1591		Johann Schmidlap of Nuremberg publishes an informative book on pyrotechnical displays with fireworks.[g]	
1592			Japanes Toyotomi Hideyoshi's first invasion of Korea: Admiral Yi Hsuan Chen of the Korean navy wins a great victory at Hsien San Island by burning the enemy's fleet with fire arrows and fire barrels.[h]
1593			During the war between Thailand and Cambodia, signal

Table II (continued)

Year	China	India, Europe	Eastern Asia
			rockets are said to be used by the Siamese at the fall of Lovec.[i]
1597			Toyotomi Hideyoshi's second invasion of Korea: the joint forces of Korea and China defeat the Japanese fleet on Luliang Sea again by firing fire arrows and fire barrels.[j]

[a] Ley (see Ref. 26), pp. 57-58.
[b] Winter (see Ref. 45).
[c] Maxwell (see Ref. 57).
[d] Winter, op. cit. (Ref. 45).
[e] Ch'i Chi-Kuang (see Ref. 85).
[f] Winter, op. cit. (Ref. 45).
[g] Emme (see Ref. 27), p. 68.
[h] Yi (see Ref. 64), p. 335.
[i] Winter, op. cit. (Ref. 45).
[j] Yi, op. cit. (Ref. 64), p. 341.

Table II (continued)

F. Development in China and the Outer World: 17th Century

Year	China	India, Europe
1612		Battle of Daultatabad (India): a Maratha rocket strikes one of Jahangir's officers and kills him.[a]
1619	Mao Yüan-I completes his Wu Pei Chih (Treatise on Armament Technology), a comprehensive book of 240 volumes with 16 volumes devoted to gunpowder, firearms, and rockets.[b]	
1635		John Babington of England publishes his book, Pyrotechnia, containing illustrations of rocket charging tools, and descriptions of rocket manufacturing.[c]
1637	Sung Ying-Hsing completes his book, Tien-Kung K'ai-Wu (The Exploitation of the Work of Nature), with one chapter devoted to gunpowder and firearms.[d]	
1643	Chiao Hsü completes his book, Tse K'o Lu (Elements of Fire Warfare), in cooperation with Schall von Bell (German).[e]	

Table II (continued)

Year	China	India, Europe
1650		Kosmierz Siemienowicz, a Polish artillery expert, publishes his book on rocketry, entitled Artis Magne Artillerias, which contains many drawings showing various rocket designs.[f]
1957		In the siege of the Fort of Bidar (India) by Aurangzeb, its final downfall is attributed to the explosion of its chamber of grenades caused by a rocket hit.[g]
1658		During the war between Aurangzeb and his brother, Dara Shukoh, both sides use rockets at Samugarh, but Dara's seat on a riding elephant is hit by a missile and he is unseated, causing a panic which leads to his defeat.[h]
1659		In the naval encounter at Dogatichi, Mughal Prince Shah Shuja's ships are struck with rockets by the Imperial fleet, causing the masts to fall, and killing a number of sailors and soldiers.[i]
1687		Isaac Newton publishes his Third Law: the Law of Reaction, in his monumental book, the Principia.

[a] Winter (see Ref. 45).
[b] Mao Yuan-I (see Ref. 86).
[c] Maxwell (see Ref. 57).
[d] Sung Ying-Hsing (see Ref. 87).
[e] Chiao Hsu (see Ref. 88).
[f] Emme (see Ref. 27), p. 68.
[g] Winter, op. cit. (Ref. 45).
[h] Ibid.
[i] Ibid.

Bibliography

A. Chinese Literature (Chronological Order)

1. Liu An 劉安, Huai Nan Tzu 《淮南子》 (The Book of the Prince of Huai-Nan), c. -120.
2. Wu Pu 吳普, Shen Nung Pen Ts'ao Ching 《神農本草經》 (Classical Pharmacopoeia of the Heavenly Husbandman) (ed. with notes by Sun Shing-Yen 孫星衍 et al., 1799), +2nd century.
3. Wei Po-Yang 魏伯陽, Chou-I Ts'an T'ung Ch'i 《周易叄同契》 (The Kinship of the Three and the Book of Changes), +142.
4. Pan Ku 班固, Ch'ien Han Shu 《前漢書》 (History of the Earlier Han Dynasty), c. +100.
5. Chang Chung-Ching 張仲景, Chin-Kuei Yao-Lüeh 《金匱要畧》 (The Essentials in the Golden Chest), 180.
6. Ko Hung 葛洪, Pao P'u Tzu 《抱朴子》 (Book of the Preservation of Solidarity Master), c. 320.
7. Fan Yeh 范曄, Hou Han Shu 《後漢書》 (History of the Later Han Dynasty), 450.
8. Sun Ssu-Mo 孫思邈, Pei Chi Ch'ien Chin Fang 《備急千金方》 (A Thousand Golden Remedies for Emergency Uses), 7th century.
9. Chin Shih Pu Wu Chiu Shu Chüeh 《金石簿五九數訣》 (Explanation of the Inventory of Metals and Minerals According to the Numbers Five and Nine), 670 (author unknown).
10. Chao Nai-An 趙耐安, Ch'ien Hung Chia Keng Chih Pao Chi Ch'eng 《鉛汞甲庚至寶集成》 (Complete Compendium on the Perfected Treasure of Lead, Mercury, Wood and Metal), 808.
11. Meng Yao-Fu 孟要甫, Chu Chia Shen P'in Tan Fa 《諸家神品丹法》 (Methods of Various Schools for Magical Elixir Preparations), 10th century.

12. Chen Yüan Miao Tao Yao-Lüeh 《眞元妙道要略》(Classified Essentials of the Mysterious Tao of the True Origin), attributed to Cheng Shih-Yüan 鄭思遠, 3rd century.

13. Liu Hsu 劉昫, Chiu T'ang Shu 《舊唐書》(Old History of the T'ang Dynasty), 945.

14. Tseng Kung-Liang 曾公亮 and Ting Tu 丁度, Wu Ching Tsung Yao 《武經總要》(Collection of the Most Important Military Techniques), Vols. 11-13, 1044.

15. Sung Ming-Chiu 宋敏求, Tung-Ching Chih 《東京志》(About the Eastern Capital), early 11th century.

16. Shih Liang-Mou 石良茂, Pei Jung Yeh-Hua 《避戎夜話》(Some Night Talks During My Hiding from the War), 1130.

17. Yang Wan-Li 楊萬里, Hai Chiu Fu Hou Hsu, Ch'eng-Chai Chi 《海鰌賦後序，誠齋集》(Epilogue of the Ode of the Sea-eel, The Collected Work of Yang Ch'eng-Chai), Vol. 44, 1170.

18. Hsu Meng-Hsin 徐夢莘, San Tsao Pei Ming Huai Pien 《三朝北盟會編》(Collection of Documents on the Treaty with the North in Three Reigns), 1194.

19. Chou Mi 周密, Chi-Tung Yeh-Yü 《齊東野語》(Country Talks of Eastern Chi), Vol. 11, 1269.

20. Chou Mi 周密, Wu-Ling Chiu-Shih 《武林舊事》(Old Anecdotes of Wu Ling), 1269.

21. Chou Ta-Kuan 周達觀, Chen-La Fen-T'u Chi 《眞臘風土記》(Life and Custom in Chen-Lau [Cambodia]), 1297.

22. Liu Ming-Chung 劉敏中, Ping Sung Lu 《平宋錄》(The Conquer of Sung), 1304.

23. T'o-T'o 脫脫 and Ouyang Hsüan 歐陽玄, Sung Shih 《宋史》(History of the Sung Dynasty), 1345.

24. T'o-T'o 脫脫, Chin Shih 《金史》(History of the Chin Dynasty), 1345.

25. Sung Lien 宋濂, Yüan Shih《元史》(History of the Yuan Dynasty), 1369.

26. Tien Ju-Ch'eng 田汝成, Yen Chao Chi Wen《炎徼記聞》(Memories on the Southern Border), 1558.

27. Ch'i Chi-Kuang 戚繼光, Chi Hsiao Hsin Shu《紀效新書》(A New Treatise on Military and Naval Efficiency), Vols. 1, 6 and 15, 1575.

28. Ming Hui Tien《明會典》(Collection of Archives of the Ming Dynasty), compiled by Imperial Order, Vol. 156, 1587.

29. Mao Yüan-I 茅元儀, Wu Pei Chih《武備志》(Treatise on Armament Technology), Vols. 126-129, 1628.

30. Chang Hsüan 張萱, Hsi-Yüan Wen-Chien-Lu《西園聞見錄》(Seen and Heard in the West Garden), Vol. 73, 1582.

31. Sung Ying-Hsing 宋應星, T'ien-Kung K'ai Wu《天工開物》(The Exploitation of the Works of Nature), Vol. 15, 1637 (English translation by E-Tu Zen Sun and Shiou-Chuan Sun, the Pennsylvania State University Press, University Park and London, 1966).

32. Chiao Hsü 焦勗, Tse K'o Lu《則克錄》(Elements of Fire Warfare), 1643.

33. Chang T'ing-Yü 張廷玉, Ming Shih《明史》(History of the Ming Dynasty), 1739, edited by Yang Chia-Lu 楊家駱 (Ting-Wen Book Co., Taipei, 1975).

34. Kuo Shao-Ming 柯劭忞, Hsin Yüan Shih《新元史》(New History of the Yuan Dynasty), 1932.

35. Tu Chi 屠寄, Meng-Wu-Erh Shih-Chi《蒙兀兒史記》(History of the Mongols), 1934.

36. Feng Chia-Sheng 馮家昇, Huo-Yao Chi Ch'i Hsi-Ch'uan《火藥及其西傳》(Origin of Gunpowder and Its Western Spread), Shanghai, 1954.

37. Chang Hsing-Lang 張星烺, Chung Hsi Chao-Tung Shih-Liao Hui-Pien《中西交通史料彙編》(The Materials for a History of Sino-Foreign Relations) (World Book Co., Taipei, 1962).

38. Weng Tung-Wen 翁同文, 'Chen Yüan Miao Tao Yao-Lüeh ti Ch'eng-Shu Shih-Tai Chi Hsiang-Kuan ti Huo-Yao-Shih Wen-T'i'〈眞元妙道要畧的成書時代及相關的火藥史問題〉, 'The Dating of Chen Yüan Miao-Tao Yao-Lüeh and Gunpowder, Nanyang University Journal, No. 5, 33-80 (1972).

39. Chen Kuo-Fu 陳國符, Tao Ts'ang Yüan Liu K'ao《道藏源流考》(Researches in the History of the Taoist Patrologies), (Ming-Wen Book Co., Taipei, 1975).

40. Li Ch'iao-P'ing 李喬萍, Chung-Kuo Huo Hsueu Shih《中國化學史》History of Chemistry in China (Commercial Press, Taipei, 1975).

41. Chao Tieh-Han 趙鐵寒, The Invention of Gunpowder (National Institute for Compilation and Translation, Taipei, 1978).

42. Cheng Hsüeh-Chia 鄭學稼, Jih-Pen Shih《日本史》(History of Japan) (Li-Ming Book Co., Taipei, 1977), Vol. 3.

43. Wang Kung-Chi 汪公紀, Jih-Pen Shih-Hua, Chung-Ku Pien《日本史話，中古篇》, Epic of Japanese History, Medieval Ages (Linking Publishing Co., Taipei, 1982).

44. Lung Ts'un-Ni 龍村倪, 'Fu-Huo-Liu-Huang-Fa Ying Kuei Sun Chen Jen, Yi Chang Yi Chih Tsui Ku Huo-Yao Hsiang Fang Pien'〈伏火硫黃法應歸孫眞人，一張已知最古火藥詳方辨〉('On the Earliest Known Gunpowder Recipe, A Method of Subduing Sulphur by Fire, by Sun Chen-Jen'), Science Monthly (Taipei), 13 (No. 11), 64-70 (1982).

45. Liu Kwang-Ting 劉廣定, 'Tsai-T'an Fu-Huo-Liu-Huang-Fa ti Tso-Che Wen-T'i'〈再談伏火硫黃法的作者問題〉

('Rediscussion on the Author of the Method of Subduing Sulphur by Fire') Science Monthly (Taipei), 13 (No. 11), 70-72 (1982).

46. Chen Kuo-Fu 陳國符, Tao-Ts'ang Yüan-Liu Hsü-K'ao《道藏源流續考》(Continued Research on the History of the Taoist Patrologies) (Ming-Wen Book Co., Taipei, 1983).

B. In English and Other Languages (Chronological Order)

1. D'Hosson, Histoire des Mongoles (Amsterdam, 1852, 4 vols, Chinese translation by Feng Chen-Chün 馮承鈞 (Commercial Press, Taipei, 1934).
2. Henry Yule, The Book of Ser Marco Polo the Venetian (Murray, London, 1903), Chinese translation by Chang Hsing-Lang 張星烺 (Commercial Press, Taipei, 1965).
3. Berthold Laufer, Sino-Iranica: Chinese Contribution to the History of Civilization in Ancient Iran (Field Museum of Natural History Publications, 1914), Anthropological Series, Vol. 15, No. 31, Chinese translation by Tu Cheng-Sheng 杜正勝 (Chung-Hua Book Co., Taipei, 1975).
4. P. K. Gode, Use of Guns and Gunpowder in India from A.D. 1400 Onwards (Karnatak Publishing House, 1939), A Volume of Indian and Iranian Studies, p. 29.
5. L. C. Goodrich, and C. S. Feng, 'The Early Development of Firearms in China,' ISIS, 36, 114-123 (1946).
6. Wang Ling 王鈴, 'On the Invention and Use of Gunpowder and Firearms in China,' ISIS, 37, 160-178 (1947).
7. J. Needham, Science and Civilisation in China (Cambridge University Press, 1954), Vol. I, Chapter 7.
8. J. Humphries, Rockets and Guided Missiles (Macmillan, 1956).
9. G. P. Sutton, Rocket Propulsion Elements (Wiley, New York, 1956).

10. J. R. Partington, A History of Greek Fire and Gunpowder (W. Heffer and Sons, Cambridge, 1960), p. 211.

11. H. Arakawa, 'Rockets Fired by the Mongol Forces in 1274 A.D.,' Proceedings of the 2nd International Symposium on Rockets and Astronautics (Tokyo, 1960), pp. 330-331.

12. Yi Pyong-Po 李丙燾, History of Korea《韓國史大觀》 (Chinese translation from Korean by Hsü Yü-Ch'eng 許宇成 (Cheng-Chung Book Co., Taipei, 1961).

13. E. M. Emme, A History of Space Flight (Holt, Rinehart, and Winston, NY, 1965).

14. N. Sivin, Chinese Alchemy: Preliminary Studies (Harvard University Press, Cambridge, Mass., 1968).

15. Willy Ley, Rockets, Missiles, and Men in Space (Viking, New York, 1968).

16. F. H. Winter, 'On the Origin of Rockets,' Chemistry, 49 (No. 2), 8-12 (1976).

17. F. H. Winter, 'The Genesis of the Rocket in China and Its Spread to the East and West,' XXXth Congress, International Astronomical Federation (Munchen, 1979).

18. F. H. Winter, 'Rocketry in India from Earliest Times to the 19th Century,' Journal of the British Interplanetary Society, 32, 467-471 (1979).

19. J. Needham, 'The Guns of Khaifeng-fu,' The Time Library Supplement (Times Newspapers Ltd., London, 1980), No. 4007.

20. Sun Fang Toh 孫方鐸, 'Rockets and Rocket Propulsion Devices in Ancient China' (14th IAA International Symposium on History of Astronautics, 31st IAF Congress, Tokyo, 1980) Journal of Astronautical Sciences, XXIX (No. 3), 289-305 (1981).

21. Joseph Needham, 'The Epic of Gunpowder and Firearms: Developing from Alchemy,' Science in Traditional China: A Comparative Perspective (Linking Publ. Co., Taipei, 1982).

22. W. R. Maxwell, 'The Early History of Rockets,' Journal of the British Interplanetary Society, 35, 176-184 (1982).

23. F. H. Winter, 'A New Look at Early Chinese Rocketry, 1200-1900,' Journal of the British Interplanetary Society, 35, 522-529 (1982).

24. J. Pan, 'On the Origin of Rockets,' Nature (Beijing), 3, 173-184 (1984).

25. Sun Fang Toh 孫方鐸 , 'On the Early Rocket Weapons in China' (18th IAA International Symposium on the History of Astronautics, 35th IAF Congress, Laussanne, Switzerland, 1984), Paper No. IAA-84-258.

26. Y. S. Chae, 'A Study of Early Korean Rockets (1377-1600),' Journal of the British Interplanetary Society, 38, 169-178 (1985).

CHRONOLOGY OF CHINESE HISTORY

Palaeolithic Populations	c. -600000 to c. -8000
Neolithic Societies	c. -8000 to c. -2000
Hsia 夏 Kingdom	c. -2100 to c. -1650
Shang 商 Dynasty	c. -1650 to c. -1111
Chou 周 Dynasty	c. -1120 to -221
Western Chou 周 (c. -1122 to -771)	
Ch'un Ch'iu 春秋 (-770 to -476) (Spring and Autumn Period)	
Chan Kuo 戰國 (-475 to -221) (Warring States Period)	
Ch'in 秦 Dynasty	-221 to -207
Han 漢 Dynasty	-206 to +220
Western Han 漢 (-206 to +8)	
Hsin 新 Interregnum (+9 to +24)	
Eastern Han 漢 (+25 to +220)	
San Kuo 三國 (Three Kingdoms)	220 to 265
Wei 魏 (220 to 265)	
Shu (Han) 蜀漢 (221 to 264)	
Wu 吳 (222 to 280)	
Chin 晉 Dynasty	265 to 420
Western Chin 晉 (265 to 316)	
Eastern Chin 晉 (217 to 420)	

Nan-Pei Ch'ao 南北朝 (Southern and Northern Dynasties)	420 to 589
Southern Dynasty (420 to 589)	
(Liu) Sung 劉宋 (420 to 479)	
Ch'i 齊 (479 to 502)	
Liang 梁 (502 to 557)	
Ch'en 陳 (557 to 589)	
Northern Dynasty (386 to 581)	
Northern Wei 魏 (386 to 534)	
Eastern Wei 魏 (534 to 543)	
Western Wei 魏 (535 to 554)	
Northern Ch'i 齊 (550 to 577)	
Northern Chou 周 (557 to 581)	
Sui 隋 Dynasty	581 to 618
T'ang 唐 Dynasty	618 to 907
Wu Tai 五代 (Five Dynasties)	907 to 960
Hou Liang 後梁 (907 to 923)	
Hou T'ang 後唐 (923 to 936)	
Hou Chin 後晋 (936 to 946)	
Hou Han 後漢 (947 to 950)	
Hou Chou 後周 (951 to 960)	
Sung 宋 Dynasty	960 to 1279
Northern Sung 北宋 (960 to 1127)	
Liao 遼 (916 to 1125)	
Hsi Hsia 西夏 (990 to 1227)	
Chin 金 (1115 to 1234)	
Southern Sung 宋 (1127 to 1279)	
Mongol 蒙古 (1206 to 1271)	

Yuan 元 Dynasty	1271 to 1368
Ming 明 Dynasty	1368 to 1644
Ch'ing 淸 Dynasty	1644 to 1911

INDEX